NUCLEAR DISASTER IN THE URALS

NUCLEAR DISASTER IN THE URALS

Zhores A. Medvedev

Translated by George Saunders

W · W · NORTON & COMPANY
NEW YORK

Published simultaneously in Canada by George J. McLeod Limited, Toronto.
Printed in the United States of America.

First Edition

Library of Congress Cataloging in Publication Data

Medvedev, Zhores Aleksandrovich.
Nuclear disaster in the Urals.

Bibliography: p.
Includes index.
1. Radioactive pollution—Russian Republic—Ural Mountain region. 2. Radioactive waste disposal—Russian Republic—Ural Mountain region—Accidents.
I. Title.
TD196.R3M42 1979 574.5'264 79-11133
ISBN 0-393-01219-0

Contents

Translator's Note on Russian Spellings

The Library of Congress transliteration system has been used for the many references to scientific titles in this book, to assist readers who might wish to check these sources. Personal and place names in the text generally conform to this system as well, but *-sky* is used rather than *-skii* (e.g., Rovinsky and Kamensk-Uralsky, rather than Rovinskii or Kamensk-Uralskii) and the apostrophe for the soft-sign is likewise omitted in the text (e.g., Ilenko rather than Il'enko, and Uralsky, not Ural'sky). Certain standard spellings, more familiar in English, are also used, such as Chelyabinsk and Novaya Zemlya (with the -ya- rather than -ia-). Many of the excerpts reproduced in this book from English-language sources use spellings that differ from this system. Although such spelling inconsistencies may be distracting, they have been kept, to avoid making changes in quoted material.

NUCLEAR DISASTER IN THE URALS

Chapter 1

A Big Sensation Begins

In the summer of 1976 I received a letter from an editor of the British popular-science journal *New Scientist,* dated July 28. It said in part:

> The *New Scientist* celebrates its 20th anniversary this year and we are going to have a special anniversary issue in November. The editor is looking for interesting articles about developments over the past 20 years. But he doesn't just want histories of this or that field over the period.
>
> 1976 is also, of course, the 20th anniversary of Khrushchev's famous speech to the party congress, which indicated a definite watershed in the life of the Soviet Union. It occurred to me that it might be very interesting to have an article dealing with science and scientists in the Soviet Union since 1956. I'm not thinking about a definitive, academic piece, but more an article that discusses various intellectual, cultural and political trends or developments. Ideas that come to mind include—

a) the development of a technological elite, its changing position in the hierarchy, its power, etc.;
b) the role of science and scientists in changing Soviet society, trends both for and against democratisation, inherent contradictions vis-à-vis the need for free information and the need to control society; and
c) the role of scientists in the dissident movement, both because their training encourages a certain amount of independent thought and/or because science at that time was attracting the most independent and creative minds.

Of the proposed topics I chose the last, because in 1956, as a young scientist, I had joined the struggle Soviet biologists were waging against a form of pseudo-science—first supported by Stalin and then by Khrushchev—known to the world as "Lysenkoism."

My article was entitled "Two Decades of Dissidence," and in recounting the developments of those decades, I mentioned that one of the most important episodes which brought an influential group of atomic physicists together with persecuted geneticists was the nuclear disaster in the Urals which had occurred in 1957 or 1958. This nuclear accident had contaminated more than a thousand square kilometers in the southern Urals with radioactive waste from nuclear reactors and had caused the deaths of several hundred persons. Thousands had been evacuated and hospitalized, and an extensive area in an industrially developed region had become a danger zone and would remain so for decades. A proper analysis of all the consequences of this disaster required the science of genetics, but in 1958 genetics was still regarded as "bourgeois" and "reactionary" in the USSR and was surrounded by prohibitions.

I had been living in England only since 1973 and had no idea that Western experts were uninformed about the nuclear disaster in the Urals. I wrote that the disaster was caused by the discharge into the atmosphere of an enormous quantity of radioactive waste which had been stored underground for many years.

At that very time there was a heated controversy in the British press over nuclear waste that was to be shipped in from Japan. In Sweden the issue of nuclear power plants and nuclear waste had been central in an election which brought down the Social Democratic government. In Germany, France, and other countries, a strong protest movement against nuclear power and the construction of nuclear plants had developed. The question of our planet's dwindling energy reserves had become a hotly debated topic. Under these circumstances there was a strong reaction to my brief description of the Urals disaster of nearly twenty years earlier. My article was published on November 4, 1976 (1). News stories about it were printed in almost all the leading Western newspapers, and there was considerable television coverage as well. Experts in Britain, the United States, France, and many other countries responded by denying my story and stating that such a thing was technically impossible. The earliest and most sharply worded rebuttal to my story came from the chairman of the United Kingdom Atomic Energy Authority, Sir John Hill. His interview with the Press Association published in the *London Times* on November 8, 1976, was reprinted in many European and American papers. He declared in the most high-handed fashion that my story was "rubbish." Since this challenge to my veracity was what prompted me to do my subsequent research on the subject, I consider it appropriate to cite the text of the interview just as it appeared in the *Times*.

> Soviet Nuclear Disaster Discounted
>
> Claims by a dissident Soviet scientist that hundreds of people died in a nuclear catastrophe in the Soviet Union in 1958 were dismissed as "pure science fiction" yesterday by Sir John Hill, chairman of the United Kingdom Atomic Energy Authority.
>
> In an interview with the Press Association, Sir John described the allegations of Dr. Zhores Medvedev as "rub-

bish" and added: "I think this is a figment of the imagination."

Dr. Medvedyev, a biochemist, claimed in an article in the *New Scientist* that nuclear waste which was buried near the surface in the Urals blew up "like a volcano" in 1958. The resulting radioactive cloud spread across hundreds of miles and thousands of people became afflicted by radiation sickness.

Sir John said that while the Russians probably did bury low-level nuclear waste, as did Britain and other countries, "this sort of waste has a very, very low activity and could not possibly give that sort of explosion."

Even if the Russians buried high-level waste—and Sir John did not believe they did, as they followed safety standards similar to those in other countries—"it could not give that sort of explosion, nuclear or thermal."

The same report, somewhat abbreviated, appeared in the *New York Times* on November 8, in the form of a Reuters dispatch.

In the United States, not only nuclear experts commented on my article; the Central Intelligence Agency did so too. The CIA had its own version of the accident, although, of course, the sources were not given. The following news story appeared in the *Denver Post, Los Angeles Times,* and other newspapers on November 10 and 11, 1976.

N-Reactor Blast of '50s Described

LOS ANGELES—American intelligence experts said Tuesday that a major nuclear accident in the Soviet Union nearly two decades ago involved a reactor that went out of control, not an explosion of atomic waste as an exiled Soviet scientist asserted last week.

Two intelligence sources told the Los Angeles Times independently that the mishap occurred in late 1957 or early 1958 and involved a plutonium-production reactor at a Soviet nuclear-weapons complex located several hundred miles northeast of the Caspian Sea near the southermost Ural Mountains.

One official said the accident, which intelligence agencies detected shortly after it occurred, was the only event in the history of the Soviet nuclear program that could accurately be described as a "disaster."

Moreover, sources indicated that it involved reactor technology only distantly related to present-day nuclear power plants and that the accident's relevance to the safety of civilian nuclear power today is probably minor. . . .

"Hundreds of deaths and thousands of injuries is hard to believe," one analyst said. "We've had various numbers on this from time to time, and there were probably some deaths in the immediate area (of the reactor), but it's very hard to verify."

He added, however, that "there was no evidence that the accident put a crimp in their plutonium production or that it gave them second thoughts about their weapons programs. They just piled dirt over the reactor—buried it that way—and went about their business."

In the European press (*International Herald Tribune, Guardian,* etc.), the same story was printed on November 12.

The name of the intelligence "analyst" was not reported, but apparently he was not "intelligent" enough to make the connection between this event and the fact that precisely in early 1958 Khrushchev suddenly announced unilateral suspension of atomic weapons tests by the Soviet Union. The "clean-up of a small amount of contamination in a reactor" actually took quite a few months. An important center of nuclear weapons production had to be restored.

Subsequently, in October 1958, despite the protests of a number of Soviet atomic physicists, Khrushchev announced the resumption of atmospheric testing with a major series of new tests.

The American and British intelligence agencies undoubtedly

have many different ways of observing developments in the Soviet nuclear industry. This is what gives authoritative figures and leaders of the John Hill type the confidence to assert that they have "a pretty good idea of what is happening in other countries." One weakness of the intelligence agencies, however, is that they are preoccupied with hunting for *secret* information. They are by no means able in every case to extract the necessary information from publicly accessible scientific sources.

Today's "analysts," who receive such a vast amount of information from the various technical instruments of modern intelligence such as computers, artificial satellites, and other means of observation, as well as from informers and defectors, are often unable to make thoroughgoing and effective use of information open to the public. In this book, I propose to give these analysts and experts a small lesson in scientific detective work. The numerous sources which are cited and utilized in the present work are not secret; they are published in ordinary scientific periodicals. The real history of the Urals nuclear disaster is recorded in the omissions, distortions, falsifications, and anomalies that appear in these published sources. The investigator who has sufficient experience working with radioactive isotopes will not find it difficult to understand and fill in the gaps. As a result of the additional material I published on this question, the CIA was forced to "declassify" and authorize for publication certain documents which I will comment on in a later chapter of this book.

Chapter 2

The Sensation Continues

My definition of a "sensation" is an event that makes major headlines on all the front pages of leading Western newspapers. In this sense, it is stretching things a bit to call my first article in the *New Scientist* a big sensation; as far as I know, it made the front page of only one paper, the London *Observer* of November 7, 1976. Other newspapers reported the story somewhere on page 3 or 4. In order to create more of a sensation, some newspapers began to construct various hypotheses as to what exactly provoked Medvedev to make the news of the disaster public at the end of 1976, almost twenty years after the event. The *Guardian* went the farthest with such speculations. Anthony Tucker, its science correspondent, without contacting me, wrote the following in a November 8 article with the heading "Russian Reveals Nuclear Tragedy":

> Medvedev was not available yesterday to explain why he had waited more than a year before telling the story.

The scientists who know him confirmed that he is a highly political man whose motive for revealing the disaster now may well be to draw further attention to British plans to build a large nuclear waste treatment plant at Windscale in Cumbria.

On the next day, I called the newspaper's offices and denied this fabrication, explaining that I had only heard about the British plans as a result of the uproar over my article. Other observers likewise attempted to link my description of the Urals disaster with the controversy in Britain and the rest of Europe over the storage of nuclear waste. This showed that none of them had read my original article in the *New Scientist* and were only commenting on distorted newspaper accounts.

Newspaper sensations do not last very long. I and many other readers of the daily press soon began to forget about the Urals disaster, especially since the very existence of that event had been officially called into question. But, unexpectedly, a month after my article the Urals explosion became a genuine sensation, making big headlines on the front pages of almost every Western newspaper. In the *Evening Standard* of December 7, a sensational story appeared under the headline "The Town That Died for Hundreds of Years." The *Daily Telegraph* had a front-page headline the following morning entitled "Refugee Saw Soviet Atom Devastation." A map of the Urals was also published with an arrow pointing to the likely site of the disaster. The same day, the London *Times* carried a front-page story in a calmer vein, "Urals Nuclear Disaster Described by Eyewitness."

All of these sensational reports came from Jerusalem, where the *Jerusalem Post* had printed a short letter in its Letters to the Editor section from Prof. Lev Tumerman, an émigré from the USSR, who had passed through the contaminated radioactive zone in 1960, traveling by motor vehicle between the two largest cities in the Urals, Sverdlovsk and Chelyabinsk. Professor Tumerman, who emigrated to Israel in 1972, had become an advocate of nuclear-power development in that country, contending that Israel's lack of its own energy resources made the construction of nuclear power plants a vital necessity. Disturbed by the American intelligence reports that the Urals disaster had been a reactor accident, Tumerman decided to stress the fact that, according to all the accounts he heard from people he talked with during his trip, both scientists and ordinary people, the disaster was caused by *negligence and carelessness in the storage of nuclear waste.* I cite the original text of his letter because what was printed in the following two days in dozens of different papers or reported over the radio was frequently marred by the distortions or embellishments of the reporters who were constantly calling Israel to obtain "exclusive" interviews with Professor Tumerman.

Soviet Nuclear Disaster

To the Editor of the Jerusalem Post

Sir, In order to counter reports that the major nuclear accident in the Soviet Union (November 7 and 11) was connected with nuclear power reactor malfunction, I would like to add my eye-witness account of the disaster.

In 1960 I had occasion to make a trip by car [to] a place near Cheliabinsk in the Southern Urals [from northeast of] the city of Sverdlovsk in the Northern Urals.* We began our trip shortly after midnight and reached the main highway leading from Sverdlovsk to the South at approximately 5 A.M., when it was clear enough to see the surrounding area.

* It is clear from interviews with Professor Tumerman, published later, that he started his journey, not from Chelyabinsk (as his letter first stated) but from the construction site of the Byeloyarsk nuclear power plant, about 300 kilometers east-northeast of Sverdlovsk. They reached Sverdlovsk a few hours later, at dawn, and there turned south on the highway to Chelyabinsk. The distance between Sverdlovsk and Chelyabinsk is about 180 kilometers, and so it was only about two hours drive. Not far from Chelyabinsk, at Miassovo, Tumerman attended a summer seminar on genetics organized by Professor Timofeev-Resovsky. At Byeloyarsk, Tumerman had been visiting his brother, an engineer on the construction project.

About 100 kilometres (60 miles) from Sverdlovsk a road sign warned drivers not to stop for the next 30 kilometres and to drive through at maximum speed.

On both sides of the road as far as one could see the land was "dead": no villages, no towns, only the chimneys of destroyed houses, no cultivated fields or pastures, no herds, no people . . . nothing.

The whole country around Sverdlovsk was exceedingly "hot." An enormous area, some hundreds of square kilometres, had been laid waste, rendered useless and unproductive for a very long time, tens or perhaps hundreds of years.

I was later told that this was the site of the famous "Kyshtim catastrophe" in which many hundreds of people had been killed or disabled.

I cannot say with certainty whether the accident was caused by buried nuclear waste, as Zhores Medvedev wrote in the *New Scientist* and the *Jerusalem Post* or by the explosion of a plutonium-producing plant, as intelligence sources (quoted by A.P. and the *Times*) have said. However, all people with whom I spoke—scientists as well as laymen—had no doubt that the blame lay with Soviet officialdom who were negligent and careless in storing the nuclear wastes.

(PROF.) L. TUMERMAN
Weismann Institute of Science
Rehovoth

I sent Professor Tumerman a reprint of my article from the *New Scientist,* since to all appearances he had not seen it but was simply reacting to newspaper accounts. A few days later I received a letter from him in which he wrote that he had decided to speak out precisely so that the disaster in the Urals would not be linked in peoples' minds with the nuclear reactors or nuclear power plants which he felt Israel needed. "To me, anti-nuclear agitation is especially dangerous," wrote Tumerman, "in our country, which lacks any sources of energy and is surrounded by hostile states which hold in their hands almost all the oil reserves of the world. I was afraid that the news of the nuclear disaster might be used as a weapon in the struggle against the construction of nuclear plants in Israel and sent the letter to the editors of the *Jerusalem Post* in which I described what I had seen and stressed that the disaster could not be related in any way to the functioning of a nuclear power plant. . . .

"It was a complete surprise to me that my note aroused such a frightful sensation. I was called by BBC, NBC, French television, and God knows who else; I was besieged by reporters from every press agency and newspaper, photographed, interviewed, and finally Swedish television even sent a crew here which taped an entire "show" in the gardens of our institute, featuring myself. To this day I cannot understand why my account aroused such interest. People kept asking me why I had never made this known earlier, and they obviously did not believe me when I told them I didn't think it would be of interest to anyone. . . ."

Chairman John Hill of the United Kingdom Atomic Energy Authority, however, remained unshakable in his skepticism. In a letter to the editors of the London *Times* he replied to an article by an opponent of the construction of a new series of fast breeder reactors. Hill referred again to the Urals disaster and continued to deny that it was possible, although he did so in a more courteous manner. It seems appropriate to give an excerpt here from that letter, which was published in the *Times* on December 23, 1976, since Hill's point of view has a direct relation to the materials presented in the following chapters—not so much his point of view on the technical aspects of the question as on the alleged impossibility of such *consequences* from the explosion as I and Professor Tumerman have described (the contamination of a vast territory by radioactive wastes).

Sir, Sir Martin Ryle quotes me on two occasions in his recent article on nuclear energy (*The Times,* December 14) and perhaps you would again allow me to put the record straight. I commented specifically upon the statements made in Dr. Medvedev's article in the *New Scientist* and the issues on the burial of nuclear waste. I made no observations on the state of Russia's technology in the late 1950s.

The most important passages in the article are "for many years nuclear reactor waste had been buried . . . the waste was not buried very deep. Nuclear scientists had often warned about this primitive waste disposal. . . . Suddenly there was an enormous explosion like a violent volcano. The nuclear reactions had led to an overheating in the underground burial grounds."

I do not believe the burial of nuclear waste in Russia or anywhere else could have led to an accident remotely resembling that described in the *New Scientist* article. The probability of there being any nuclear reaction is extremely remote and even accepting that remote possibility, the probability that it could have led to the consequences described are even more improbable. There may have been some other accident, but at a time when the public are concerned about the problems of nuclear waste I feel I should make it absolutely clear that in my view the burial of nuclear waste could not lead to the type of accident described. . . .

Yours faithfully,
JOHN M. HILL
11 Charles II Street, SW1
December 21

Undoubtedly, this point of view reflects the opinion of many other administrators and experts who deal with the *technical* side of nuclear power. It is also typical of Western intellectuals to doubt or fail to understand the possibility of *total censorship* and complete refusal to discuss events on such a large scale. Among intelligence agents who specialize in Soviet nuclear technology, the opinion is apparently widely held that they could never have "missed" such a crucial event. They knew that the southern Urals region was a center of the Soviet nuclear industry and the site of the first military reactors. The two largest industrial cities—Sverdlovsk and Chelyabinsk—and all the adjacent region have always been closed to foreigners. It was over these very regions that on May 1, 1960, the American U-2 spy plane piloted by Francis Gary Powers was shot down. According to the memoirs of Khrushchev published in the United States (3), another U-2 plane had flown over the Sverdlovsk region and the southern Urals not long before this event, but the surface-to-air missiles were not yet in place, and fighter planes could not reach the spy plane's altitude of 21 kilometers. Both the first and second U-2 flights in 1960 (and a great many more in the preceding years that Khrushchev tells about in his memoirs) were engaged in photographing all the districts in the Urals region, above all, the Sverdlovsk and Chelyabinsk areas. The route from Afghanistan to Norway over the Urals was the usual one for the U-2 planes for many years and therefore one would obviously expect that all the necessary information about a serious disaster in this area could be obtained from an analysis of these photographs.

Rumors and word-of-mouth reports about "some sort of disaster" in the Urals region in 1957 had been known to the CIA from the testimony of a number of émigrés, from Soviet agents who defected to foreign intelligence agencies, and from the CIA's own agents inside the Soviet Union, for example, the fairly well-informed Oleg Penkovsky. That is why the CIA commentaries about an accident in a military reactor which only required a slight "cleaning up" were given to the newspapers. The actual documents in the possession of the agency, which were published a year later, are the best rebuttal to this original "toned-down" version.

The present analysis of what happened in late 1957 (or early 1958?) in the Chelyabinsk region has by no means been written for the purpose of revealing sensational secrets that I learned when I worked in the USSR. Although I did know many details of the Urals nuclear disaster as early as 1958, the information certainly did not come from secret sources. Millions of people who lived in the Urals knew about this disaster, although most ordinary people thought the story that a nuclear waste storage site had exploded was absolutely false; they were more inclined to believe the inevitable rumors that an atomic bomb had accidentally exploded. It would have been unrealistic to expect to hide the existence of the disaster from the population of Sverdlovsk, Chelyabinsk, and other cities. The hospitals and clinics in those cities were filled with thousands of evacuated inhabitants, who were held for observation. After a time, when symptoms of radiation sickness began to appear in more distant areas, the evacuation zone was enlarged and people began to be placed not only in hospitals but also in sanatoria and "houses of rest" (vacation facilities) which were re-equipped as hospitals. Hunting and fishing were prohibited throughout the southern and central Urals and for several years the sale of meat and fish in private markets and collective farm markets was not permitted without special inspection for radioactivity.

However, in order to make a serious scientific analysis of the Urals disaster, I did not have to rely on memory alone. Personal recollections cannot, in any case, serve as objective proof. The main thing I had to do in order to draw objective conclusions about the nature and dimensions of the catastrophe and the present situation in the radioactively contaminated area was to consult the author and subject indexes of the American reference journal *Biological Abstracts,* then study the materials in Soviet publications at one of the branches of the British Library in London (Bayswater Branch) where there is a fairly complete collection of Soviet scientific magazines and books. Several photocopies of articles necessary for this work were obtained from the National Lending Library for Science and Technology. The Soviet journal *Genetika* (Genetics), which is cited here several times, comes by subscription to the library of the institute where I work. Some books and collections of articles with data of interest are in my personal library. Finally, I obtained some reprints from Soviet authors by following standard request procedures.

Besides the material in Soviet scientific publications, there is additional objective evidence which I did not use when I carried out this research. The CIA documents had not yet been published (although, as we shall see in chapter 12, much of it corroborates the evidence in the Soviet materials that I have analyzed). There is more that remains classified by the CIA, for example, the evidence in the microfilm photographs taken by U-2 planes over a period of years.

One piece of objective evidence, supposedly contradicting my account, does not in fact have any bearing. After my first article, some British experts did a re-analysis of the radioactivity on air filters from 1957 to 1958. In the *New Scientist* of December 23–30, 1976, we find that the United Kingdom Atomic Energy Authority, after "checking through their filters for the period November 1957 to February 1958 . . . found nothing abnormal in the total beta activity in the U.K. atmosphere. They conclude that it could not have been a nuclear explosion that released radioactive fallout into the atmosphere." I don't know how much time this duplication of effort took or what the need for it was in general. My article reported an explosion of nuclear waste, which could only contaminate a local area, although a rather extensive one; it could not contaminate the upper reaches of the atmosphere or the stratosphere, which can carry radioactivity around the world.

In analyzing the material in Soviet publications, I have tried to make it comprehensible for the general reader and interesting for radiobiologists, radioecologists, and geneticists, not only foreign but Russian as well. The Soviet authors of the works analyzed here will undoubtedly find it useful to learn how ineffective their attempts to hide, change, or even falsify their data proved to be—data that had no chance of passing the censorship if presented forthrightly.

In writing this book, I also had in mind those who relied on the meager data of the secret service as grounds for calling my first article "rubbish," "science fiction," and "a figment of the imagination." But above all my aim has been to help those who are concerned with stopping the nuclear contamination of the environment in which humanity must be able to live for millions of years to come. Politicians plan in terms of two or three decades when they make their decisions. Nuclear energy specialists sometimes weigh their decisions in terms of several centuries. Biologists and geneticists, among whom I count myself, think about the future from the point of view of evolution, constructing future models with reference to millions of generations.

Chapter 3

The Urals Disaster

In my second article in the *New Scientist* (2), in 1977, I wrote that I had first learned about the nuclear disaster in the Chelyabinsk region from Professor Vsevolod Klechkovsky, director of the department of agrochemistry and biochemistry at the Timiriazev Agricultural Academy in Moscow. At that time I was working in his department as a senior research scientist in the laboratory of biochemistry. Klechkovsky was a leading expert on the use of radioactive isotopes and radiation in research on plants and soil. He was a consultant to the State Committee for Atomic Energy under the USSR Council of Ministers and took part in the work of many other government commissions and advisory bodies on atomic energy. In 1958 he was given the job of organizing an experimental station to study the effects of radioactive contamination on plants and animals in the Chelyabinsk region. I was on very friendly terms with Klechkovsky; not long before, he had recommended my paper on selective autoradiography for presentation as a report to a UNESCO conference, the First International conference on the use of Radioisotopes for Scientific Research, held in Paris in September 1957. Klechkovsky and I were members of the Soviet delegation and shared a room for about two weeks in a hotel on the banks of the Seine.* This was my first trip outside the Soviet Union, and my last until 1973. (In 1973 I was given permission to travel to London and, while there, was deprived of my Soviet citizenship and denied the right to return to my homeland.)

In 1958 Klechkovsky began to select researchers for the Chelyabinsk experimental station. Since by then I had acquired a fairly solid working knowledge of radioactive isotopes, he offered me quite an attractive position as head of one of the proposed laboratories. Work at the station would not have required compulsory relocation to the Chelyabinsk region; I would be allowed to winter in Moscow. However, anything having to do with research in this area was considered top secret. This meant I could not publish any of my findings and would have to sign a statement agreeing not to meet or correspond with foreigners, not to travel abroad, etc. Even my contacts with citizens of my own country were subject to surveillance by special departments, that is, the KGB. This prospect did not appeal to me and I refused.

However, a number of other young researchers in my department did accept posts in the Chelyabinsk region. Klechkovsky remained in Moscow to oversee his department and carry out his duties in a number of other positions. He had, however, over-all scientific supervision of the experimental station and was involved with its work until his death in 1971.

In 1958 I learned some details of the Urals disaster from Klechkovsky. At that time I was not interested in the exact date, but certain aspects of professional interest came to my atten-

* By a quirk of fate I was provided with a room in the same hotel, l'Hotel du Quai Voltaire, in early 1978 when visiting Paris to give a talk on the ecological consequences of the Urals disaster.

tion. The main detail was that it had been an explosion involving concentrated waste produced by military reactors and stored somewhere underground. The radioactive fission products which had accumulated for many years were released explosively to the surface of the earth and carried by the wind (or a snowstorm) for dozens of kilometers. The experimental station was to be located somewhere on the edge of the contaminated zone, although even at that location the level of radioactivity was several times higher than "normal." There were no large cities in the main contaminated area but there were villages and workers' settlements. Because of the suddenness of the explosion and the dispersion of the radioactivity, the levels of contamination in the various localities were only determined after some delay. The secrecy surrounding all this work also hampered timely radiation monitoring. The first seriously organized evacuation was begun after several days, and then only in the settlements closest to the site of the explosion. Subsequently, symptoms of radiation sickness began to appear in more distant areas. The necessary treatment techniques had not been adequately developed at that time. The evacuation affected several thousand persons, possibly tens of thousands, but the number who died of radiation sickness remained unknown. The number of human casualities in such cases is a highly relative question, since radiation damage, especially involving the absorption of radioactive strontium and cesium, may not appear in the form of radiation sickness or other pathologies for many months, or sometimes years and even decades. The next generation also suffers severely since it is born of parents with an increased burden of strontium-90 in their bones and therefore with a higher level of harmful effects upon the reproductive cells.

In the following years, after I had left Moscow and was working at a research institute of medical radiology in Obninsk, a town a hundred kilometers south of Moscow, I had occasion many times to hear accounts of the Urals disaster and to meet people who had worked in the contaminated zone. Long before the revelations of Professor Tumerman I knew that there were signposts along the roads between Chelyabinsk and Sverdlovsk warning of the danger of radioactivity, that drivers were urged to go at top speed and were not allowed to get out of their vehicles. Danger signs were also posted all around the zone in the woods, in open areas, and within the zone itself. The houses in the villages and settlements were destroyed not by being blown up (in which case the trees would have suffered) but by being burned down, so that the former inhabitants would not try to retrieve their contaminated belongings.

Despite the tragic nature of the disaster, the existence of such an extensive zone of contamination containing radioactive materials at different levels of concentration offered a unique opportunity for scientific research in such areas as radioecology, radiogenetics, radiobiology, and radiotoxicology. There were a great many research laboratories, institutes, and centers of various kinds in the USSR in between 1958 and 1960 concerned with the general problems of the military and peaceful uses of radioisotopes and radiation. These institutions organized experiments in small plots, in special large-size wooden boxes, in glass containers, and in small remote ponds, in order to study—under various strictly controlled experimental conditions—the spread of radioactive isotopes in the environment, their transfer from plants to animals, the absorption of various isotopes by algae in ponds, and many other questions of radiobiology, radioecology, and radiotoxicology. The sudden appearance of a vast *natural* territory contaminated by radioactivity provided thousands of researchers with totally new opportunities and unique prospects such as had never existed in any country.

But the secrecy surrounding the whole problem of the *explosion* doomed any hope of exploiting these possibilities and prospects. Ever since 1951 when I began my experimental research with radioisotopes (at first on simple patterns of isotope distribution in plants and then on the use of radioactivity to study the localization of protein and nucleic-acid synthesis in plant tissues), I found that publication of the results of such research presented tremendous difficulties even for work having nothing to do with classified matters. According to universally binding rules, any article prepared for publication had to go through a "commission" before being sent to a journal. These special "commissions" existed at all institutes, universities, and other scientific centers. Their job was to draw up a document certifying that the article in question did not "contain information of a secret character." Without such a document no publishing house, journal, or serial publication could accept an article. The censorship agency (Glavlit) dealt only with the executive editors at the publishing houses and editors-in-chief of journals. Articles could be submitted to the censor only if accompanied by a document attesting to the absence of any classified information. Without permission from the censorship no manuscript could be sent to a printshop.

At that time, it was *common knowledge* that *everything* having to do with radioisotopes and radiation was classfied. Commissions had no authority even to consider the question of publication if a scientific article contained such words as "radiation," "radioactivity," or "radioisotopes." It made no difference what the context was. Any article with such terms had to be sent for further verification to a special censorship body of the State Committee for Atomic Energy. This committee subjected every article to an additional review by experts, often for a very long time.

All of this related to scientific work carried out at nonclassified, "open" institutions. Secret laboratories, like the experimental station established in the Urals contaminated zone, had no possibility whatsoever of publishing any of their findings.

In late 1964, after the removal of Khrushchev, the Lysenko era in biology also came to an end. In a very short time, in 1965, dozens of new centers and laboratories were established for the study of genetics, radiogenetics, population genetics, radiobiology, biophysics, and many other theoretical lines of study in biology. An experimental base was needed for all of these new centers. The Institute of Ecology of the USSR Academy of Sciences was founded in Sverdlovsk—quite close to the radioactive zone in the Chelyabinsk region. At the same time, the State Committee for Atomic Energy was reorganized, its chairman Vasily Yemelyanov being pensioned off. The procedure for publishing scientific papers on nonclassified research with isotopes and radiation was also changed. They no longer had to be sent to the State Committee; a decision by a local commission was sufficient, and such commissions were reduced in number from six members to three.

For a researcher, the *publication of one's findings* is a question of major importance. Only a published work gives one a feeling of satisfaction. The desire of scientists to be recognized for their priority in certain discoveries and to have the prestige connected with publication of their own papers cannot be underestimated. There were many groups of scientists working in the secret scientific complex around the Urals disaster zone. (In addition to the experimental station directed by Klechkovsky, many other laboratories and stations had sprung up there.) More varied groups of scientists were working in the fields of ecology, radiobiology, genetics, and radiotoxicology at "open" institutes (at certain institutions under the Academy of Sciences of the USSR as a whole, at its Urals and Siberian branches, at Moscow University, Novosibirsk University, Sverdlovsk University, and many other universities). Under these conditions, it was inevitable that opportunities for collaboration arose even without special decisions by the government. Such joint research at last made it possible for research findings to appear in the scientific press. In many of the open institutes, such as the Institute of General Genetics, the Institute of Evolutionary Morphology and Ecology of Animals, the Institute of Forestry, the Institute of Soil Sciences, the department of biology at Moscow University, and the Institute of Cytology and Genetics of the

Siberian Branch of the Academy of Sciences, the commissions which decided the fate of works submitted for publication were composed of the same scientists who wanted to see their own articles and the articles of their colleagues printed. Among the co-authors, of course, were researchers at the "secret" stations. But as a rule publications were submitted in the name of some ordinary academic institute and the stations were not referred to by name. The general censorship could be satisfied (or its vigilance disarmed) if, in describing methods, details that the censorship considered sensitive were simply not mentioned. These details included the *locations* where the work was carried out, the *causes* of the radioactive contamination, the size of the *total area,* and some other specifics. There are certain obligatory standards in scientific research, especially the description of methods. These standards are invariably observed in papers on radioecology published in the United States, England, and other countries. These same standards may be found in Soviet research on radioecology done with artificial or model systems that have truly been established for experimental purposes. It was impossible to observe such standards in describing the research done in the region of the Urals disaster, as we shall see in the chapters below.

If I had had to look through all Soviet writings related to radioecology, radiobiology, or radiogenetics, I would undoubtedly still be making my way through thousands of different research papers. My situation was simplified by the fact that I knew the names of some of the scientists who had begun work in the Chelyabinsk region with Klechkovsky from 1958 to 1959. The names of these young researchers disappeared from the scientifific literature after 1958, although they had published frequently before then. (I myself had jointly published two articles with some of them during 1956 and 1957.) Their names could not be found in the author indexes of international reference journals, such as *Biological Abstracts,* or in *Letopis zhurnalynykh statei*—the comprehensive bibliographic reference publication for Soviet literature. I could not find the names of my colleagues from 1959 through 1965. Everything they did was kept in the confines of their laboratories, apparently in the form of manuscripts which were submitted to a "special section." Suddenly, in 1966 and 1967 (when they were again permitted to publish), their names began to reappear in scientific journals, but always grouped with other authors, who had "open" institutional affiliations and whose names were new to me. This greatly facilitated my search—each new name meant a new object for my investigations. I was able to find these names in the following years listed among other groups of authors, but the topics always remained the same. By using these "control names," I was able to extract material dealing with the Urals disaster from the varied mass of "radioactive" literature.

Chapter 4

Radioactive Contamination of Lakes, Water Plants, and Fish

Secrecy Surrounding the Research

Immediately after the successful testing of the first experimental reactor, the construction of large reactors for the production of plutonium was begun in the southern Urals, the first such reactor being put into operation in 1947 (see below, Chapter 13 and [67]). At that time, a radiobiological center was also established not far from the industrial complex in the Chelyabinsk region. As was the custom in the Stalin era, this secret center was not set up as a normal scientific institution would be. Instead, it arose within the inner recesses of the Ministry of State Security (the MGB). This was a "special camp," where the main work was done by Soviet prisoner-scientists and by experts deported from Germany. Of course there were also free workers in the "special center," but they worked under contracts which denied them the right to travel freely in the country or to change job locations. The research on radiobiology and genetics at this center was headed by N. V. Timofeev-Resovsky, a scientist of world renown, who had been one of the founders of the science of radiobiology. He had emigrated from the USSR in 1926 and worked in Germany, not far from Berlin. Between 1926 and 1945 he published more than a hundred papers on radiation genetics and biophysics and several books that have become classics. After the defeat of Germany, Timofeev-Resovsky was arrested and sent, in 1946, to the USSR. After that he "disappeared," and his many European and American friends and colleagues were unable to obtain any information about his fate.

But the special center in the Urals was not his first destination in the USSR. In 1946 he was sentenced as a "German spy" and sent to an ordinary prison camp in Kazakhstan. By 1947, when they began to scour the prisons and camps for all those who had anything to do with physics and radiation, Timofeev-Resovsky was already in a hospital. He would hardly have survived another month or two, though he was only forty-seven at the time. Thus he was taken to Moscow, and only after several months of treatment sent to the Urals. His wife, also a radiobiologist, was likewise removed from Germany and brought to the Urals. Timofeev-Resovsky set about organizing the first serious research center on radiobiology and radiogenetics in the USSR, bringing in some of his former associates, as well as people who had been arrested and deported from Germany earlier (for example, S. Tsarapkin, his son L. Tsarapkin, and K. G. Zimmer) and several other scientists, biophysicists, and radiobiologists, who were to be found in the various prisons and camps. (N. W. Luchnik, for example, who now heads the department of biophysics at the Institute of Medical Radiology in Obninsk, was "summoned" to work with Timofeev-Resovsky in 1947 from a prison camp in Transcaucasia.)

In 1949 the ban on research in classical genetics found its way even into the system of prison institutes. Thus Timofeev-Resovsky's group was transferred to work in the field of radio-ecology. In a short time the group worked out new methods of research on the propagation of various radioisotopes through forests, fields, and other biocenoses and through aqueous systems and laid the theoretical foundations for "radiation biogeocenology."* In over-all charge of the work of this radiobiological and radiological center was A. I. Burnazian, deputy minister of health, who at the same time held the military title of lieutenant-general.

When the radiobiological center was shifted from genetics to radioecology, a number of young specialists in agrochemistry and soil sciences had to be "recruited." In 1949 I was a fourth-year student in the faculty of agrochemistry and soil science at the agricultural academy in Moscow, and I had many friends among the graduates who in May 1949 had completed their fifth and final year of study. Students in the USSR, after receiving their diplomas, are subject to "assignment" (*raspredelenie*)—they are offered positions in various locations. For students in agrochemistry and soil sciences such positions were of course mainly in agriculture. Among the students awaiting assignment in 1949 there was a grest sensation, the meaning of which I came to understand only many years later. Among the assignments being offered were six secret positions at some top-security institution in the Urals designated solely by the phrase "P.O. Box——." Three of these positions were with the Ministry of Health and three with the Ministry of Internal Affairs. The very best students were chosen for these openings and they "disappeared" for many years. Only in 1964 did I see some of

* The term "biogeocenology" was coined by Timofeev-Resovsky. It means essentially the same as "biocenology" but was introduced to stress that the geological, geographical, geochemical, etc., elements in an environment are an inseparable part of the complex, together with the living things, the "biocenosis" proper.

these students again, when Timofeev-Resovsky moved with his group of co-workers to the Obninsk Institute of Medical Radiology.

In 1956, after Khrushchev's "secret" speech on Stalin's crimes, the status of the prison-camp scientific center headed by Timofeev-Resovsky was changed to that of an "open" laboratory and became part of the Urals Branch of the Soviet Academy of Sciences as a "biophysics laboratory." The laboratory was in Sverdlovsk, and its experimental area, in the Chelyabinsk region, at Miassovo near Lake Miass. Beginning in 1956, the laboratory staff published several dozen articles and symposium volumes on the study of radioecology.

After the Urals nuclear disaster it was natural for this large group of experienced scientists to begin work on the radioecology of the contaminated area. But this scientific center consisting of former prisoners, some of whom had never been formally rehabilitated, was not suitable for top-secret research. Therefore a parallel scientific center, a secret one, was established in the very same part of the southern and central Urals. The "open" biophysics laboratory of the Urals Branch of the Academy of Sciences from 1958 to 1966 published its works in the ordinary scientific press. The "closed" scientific unit prepared only secret reports.

I write about this paradox to illustrate the profound differences in basic methodology practiced by two scientific centers working side by side. When in 1966 and 1967 the researchers at the "secret" station began to publish their findings (in part), they discussed exactly the same scientific problems as their colleagues at the "open" center. But whereas Timofeev-Resovsky's associates always described with the greatest precision the basic conditions of their experiments, the dose of radioactivity, the time element, the ways in which the experiments were organized, where they were carried out, the climatic and other conditions, the scientists from the "secret" laboratories were oblighed to hide many things in their publications, to leave many things out, and to distort certain details. Moreover, while the biophysics laboratory worked on certain problems through model systems (artificial ponds in large glass vessels, and artificial "soil-plant" systems in large wooden box-like structures), the other group worked on the same problems under conditions whose dimensions were quite different. The Timofeev-Resovsky group also made use of natural environments and small ponds, but always with a very precise description of the timing, the quantities, and the composition of the isotopes introduced into these environments, with a full accounting of their distribution throughout the system and a complete summary of the amount of radioactivity remaining at the end of the experiment. From 1958 to 1963 Timofeev-Resovsky's laboratory studied the fate of seventeen different radioisotopes in various biocenoses. These were both long-lived and short-lived fission products. The materials that later began to be published by the secret laboratories dealt only with the distribution of strontium-90 and cesium-137. But by 1967 the behavior of these isotopes in various systems had already been studied quite well, both in Soviet experimental research and in foreign research.

In 1964, after Khrushchev's retirement, genetics finally became "legal" in the USSR; T. D. Lysenko lost all of his former influence; and the rapid establishment of genetic laboratories and centers was necessary. In this situation it was natural for Timofeev-Resovksy to return to his work in the field of radiogenetics and evolutionary genetics. In 1964 to 1965 Timofeev-Resovsky and his closest associates moved from their "exile" in the Urals to Obninsk, a scientific satellite-city of Moscow. Here Timofeev-Resovsky became the head of the genetics department in the newly established Institute of Medical Radiology. At the end of 1962, I too moved to Obninsk to set up a molecular radiobiology laboratory at the same institute. In 1965 my laboratory became part of the department of genetics and radiobiology headed by Timofeev-Resovsky.

In the days of classified research in the field of atomic energy, it was the custom to use code words in place of ordinary scientific terms even in classified reports. This practice reflected the mistrust of secretaries, messengers, and others who were included in the various types of research only for technical or financial purposes. Timofeev-Resovsky continued to use this classified jargon in conversations about the Urals disaster.

The explosion that contaminated such a large territory in the Chelyabinsk region he referred to as the "spit-up" (*plevok*) and the nuclear waste storage site as the *yushka*. This word in the Russian dialect of the Urals (according to Dahl's dictionary of the Russian language) denotes the thick grease that forms on the surface of fish soup while in preparation. Since fish soup is usually cooked in large kettles, it was in that sense that Timofeev-Resovsky used the word *yushka*. He had in mind a kettle full of a thick, concentrated, and *hot* solution of radioisotopes. In 1965 I had little interest in the problems of ecology and the storage of radioactive wastes. I was preoccupied with fundamental mechanisms of differentiation and aging and the appearance of radiation-caused somatic mutations in these processes. I refer to Timofeev-Resovsky's research in radioecology in the Urals before 1958 because he was the real founder of this branch of science in the USSR. By nothing more than his anonymous secret reports from the prison institute from 1948 to 1955, he exerted an influence on the work of many other groups. The former prisoners later found some data from their own reports in the publications of "free" scientists working in the atomic field. The imprisoned researchers produced the scientific findings, but Lieutenant-General Burnazian and others received the rewards, titles, and decorations. After 1956, Timofeev-Resovsky was able to publish his research under his own name. In a short time he and his associates published several dozen scientific papers, symposium volumes, and entire books in the field of radiation, bioecology, and biogeocenology. As examples I will cite several basic works where bibliographic references will be found for the full extent of this research (4–8). Of these, the first two were submitted as dissertations for academic degrees by Timofeev-Resovsky and his wife—although each of them at that time had more than a hundred scientific publications to their names and an international reputation. Timofeev-Resovsky was sixty-two and his wife was sixty-three. But because they had lived in Germany before their arrest, their previous record was not considered in evaluating their scientific qualifications for work in the USSR and in their formal confirmation as directors of research groups, so that they had to write and "defend" dissertations conforming to Soviet standards. Final confirmation of these academic degrees came only in 1965 after Lysenko had been removed from his leading posts.

Two Lakes Were Contaminated in 1957–1958

The formulas, conclusions, and experimental data which, from 1957 to 1963, had been produced quite clearly under strictly controlled model conditions for seventeen radioisotopes and their mixtures suddenly became the subject of one more research paper, whose author for some reason did not refer to the conclusions of Timofeev-Resovsky and his associates, although it was easy to find these conclusions in such journals as the *Doklady* (Proceedings) of the Soviet Academy of Sciences, *Botanicheskii zhurnal SSSR* (the Botanical Journal of the USSR), and *Biulleten Moskovskogo Obshchestva ispytatelei prirody* (Bulletin of the Moscow Society of Natural Scientists). I came across this work by accident while looking through the Soviet journal *Atomnaia energia* (Atomic Energy). The author, F. Ya. Rovinsky, was not among the names I knew but was mentioned in a work I will discuss further below. The title of Rovinsky's work (9) was purely theoretical, as was the research problem he dealt with. The author presented an imaginary, round, non-running-water lake with a concave bottom and thick bottom silt deposits, which were gradually absorbing a radioisotope which had been introduced (theoretically) into the body of water on one occasion. Since the isotope was retained only temporarily in the biological components of the body of water but was permanently retained in the bottom silt deposits, it was proposed that the biomass be disregarded and the lake be considered a two-component system. This dual-component approach made it possible to extract a mathematical formula for the rate of reduction in the isotope concentration in the water over time. The result was a theoretical curve showing a rapid reduction of the isotope at first and a gradual approximation to equilibrium (a plateau) approximately over the course of one year.

After the formula and the theoretical curve had been obtained, they had to be tested in natural conditions. Rovinsky did not carry out any experimental research for this purpose, but was given some already existing figures by some undisclosed source. These showed changes in radioactivity in two non-running-water lakes which had been contaminated at one time by a mixture of radioisotopes. Among these were both short-lived and long-lived isotopes, the author making specific reference only to strontium-90. Rovinsky's paper was submitted for publication in May 1964. Taking into account the time required for such a paper to be approved for the press, we must suppose that the last measurements of activity in the lake were made no later than the fall of 1963, after which the lake was covered by ice for five to six months, as is normally the case for lakes in 90 percent of Soviet territory. But measurements of radioactivity had been taken for sixty-five months, that is, were begun somewhere between 1957 and 1958.

Nowhere does the author give the absolute figures for the real concentration of radioisotopes in the water; he gives only certain relative figures and logarithms based on the initial amounts of contamination. The experimental measurements made available to the author essentially coincided with the theoretical curves. However, the description of the two lakes in the article was rather puzzling. "The experimental bodies of water were lakes of the eutrophic type; the first was 11.3 square kilometers and the second 4.5 square kilometers. The bottoms were shallow, saucer-shaped. They had very thick silt deposits which completely leveled out the original rough topography of the bottom. The shores were partly overgrown with bushes. . . . [There were] excellent conditions for development of the biomass: high summer temperatures, good penetration of the water by sunlight to a considerable depth, etc. The hydrochemical composition of the lake water is given in Table I" (9, p. 380). To judge by this table, the hydrochemical composition of the lakes was quite varied, consistent with the varied geological structure of the bottom rock. The sodium content of the water in the second lake was nine times higher than in the first, potassium five times higher, magnesium two times, and chlorine twenty times; and there was substantially more calcium in the first lake. Thus, it is hardly likely that the lakes were situated side by side.

A natural question arises, Why were *two* lakes necessary and why of such size? The problem could have been tested easily under artificial conditions of the type in Timofeev-Resovsky's experiments. If *natural* conditions were desired, small ponds could have been found (one to two hectares* in size or less). But two lakes with a total area of more than 15 square kilometers and with a large biomass would be of great value for commercial fishing. Surely villages or other settlements would be located on or near them. Why contaminate them with a mixture of isotopes?

Reference maps of the USSR represent lakes of this size on a scale of one to four million, one cm. corresponding to 40 kilometers. The part of the USSR with the most lakes is Karelia, but it is in the north and there are no "high summer temperatures" there. Among the dozens of areas in the continental part of Russia there are more lakes of all types (both running-water and non-running-water) in the Chelyabinsk region. On my map, which has the above-mentioned scale, there are about fifty of

* A hectare, in the metric system, equals 10,000 square meters, or, 2.471 acres.

them and quite a few are exactly the size indicated by Rovinsky. Lev Tumerman has said, on the basis of what he heard from Urals residents, that Kyshtym was the town nearest the site of the disaster; the road they traveled was 40 kilometers east of Kyshtym; and that whole territory is literally dotted with lakes of both the running-water and the non-running-water kind. Several of these lakes fit the picture exactly in terms of size.

But these are only my guesses. Although the dimensions of the "experimental" lakes are *geographically* significant, and such lakes have names and are included in basic reference works on the lakes of the world, Rovinsky does not give their names or their geographical locations. Nor, as we have noted, does he give the actual level of radioactivity in the water. It is highly unlikely that these lakes, containing approximately one hundred billion liters of water, were contaminated for experimental purposes; moreover, Rovinsky was given the figures for changes in radioactivity in ready-made form five years after the fact. Even to obtain experimental "tracer doses" of strontium in the water of two lakes of such size, it would have been necessary to introduce no less than 5,000 curies of strontium-90, an amount of radioactivity found only in industry, not in experimental work. Any lake on or near which a nuclear reactor or reprocessing plant was located could have been contaminated to these levels by an accidental discharge on one occasion. But in the "experiment" there were *two separate lakes* located on different geological bedrock. And they were contaminated *at the same time*. How could such a situation arise?

But all we have so far are questions and enigmas. The only thing we can conclude with certainty on the basis of Rovinsky's paper is that the contamination of the two lakes by a mixture of radioisotopes occurred in either 1957 or 1958.

As may be seen from the illustration reproduced from Rovinsky's work (Figure 1), the theoretical and experimental curves fail to correspond only during the first twelve to thirteen months; after that they are identical. The theoretical curves

Figure 1. Strontium-90 in non-running-water lakes. (Rovinsky, 1965.)

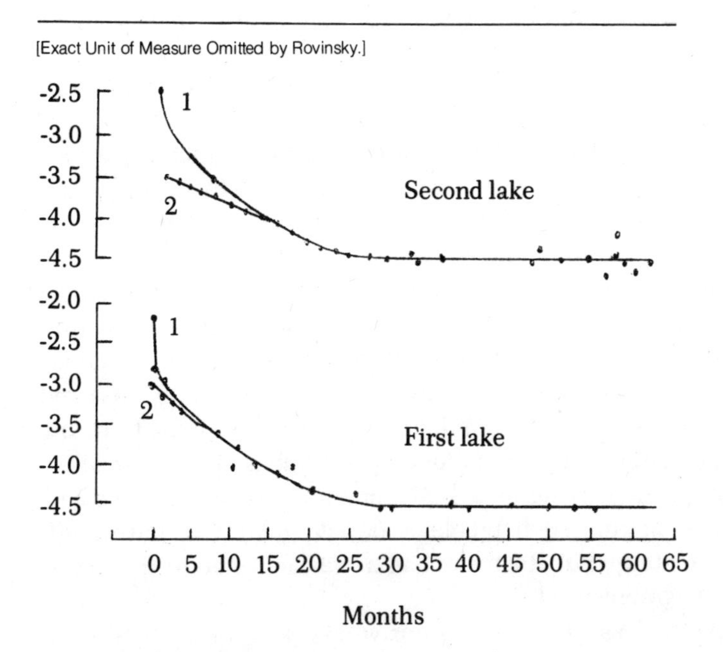

Comparison between the actual change (1) in the concentration of isotopes in the water of experimental lakes, and the estimated curve (2) for change in the concentration of strontium-90 in the water of such lakes.

calculated by Rovinsky were for the possible fate of strontium-90. In the actual lakes, as the author himself explains, the total radioactivity of the mixture of the radioisotopes was measured (but the composition of the mixture is not given). It is quite evident that during the first twelve months, short-lived radioisotopes constituted no less than 40 percent of the total contamination in the lakes; these disappeared at about the time that the theoretical and "experimental" curves began to coincide. A mixture of radioactive isotopes with a 60 percent predominance of long-lived fission products (primarily strontium) is typical of reactor waste after it has been stored for a certain length of time, or of a mixture of "old" and fresh waste with the "old" waste predominating. But how and by what these lakes were actually contaminated in 1957 or 1958 the author does not indicate, and the consideration we have indicated above must serve only as a hypothesis.

The lakes referred to in Rovinsky's article contained a great deal of oxygen and were rich in biomass. Without any doubt there were fish in these lakes. In Siberia and the Urals there is commercial fishing in lakes even smaller than these, and in the European part of the Soviet Union fish are even stocked in ponds. Therefore, it would be quite natural for the two non-running-water lakes referred to be Rovinsky to be used for serious study of the radioecology of fish (the study of food chains) and water plants (their accumulation of various isotopes from the water). These would be unique opportunities. In the small artificial ponds serving as a basis for the work of the Timofeev-Resovsky group, such problems of *fish radioecology* were difficult to study, but Rovinsky had a radioactive aqueous system on a geographic scale.

Nevertheless, the most extensive research in the USSR on certain problems of radioecology involving fish and water plants under natural environmental conditions was carried out from 1969 to 1970 in a running-water lake, which is much less appropriate for these purposes. Again, neither the name of this lake nor its geographical location has been indicated in published articles.

The key research, which upon analysis raises many interesting questions, was published in two articles by A. I. Ilenko (10, 11) and in his book (12) which generalizes not only on lake bioecology but also on zooecology in areas unquestionably located near the lake being studied. (This is evident from the fact that the water was sampled and the fish were caught during the same time that the animals and birds were trapped or shot.)

In 1969 Ilenko published a long survey article "The Radioecology of Fresh-Water Fish" (13). (It was submitted for publication in early 1968.) But this survey had no data from the research he carried out at what we will call Lake X (to be discussed further below). It must be supposed that this survey was the fruit of methodological and theoretical preparations which preceded the beginning of the main experiments. The first, very brief article on the study of Lake X was not published until 1970 (10). It was entitled "The Accumulation of Strontium-90 and Cesium-137 in Fresh-Water Fish," and the brief description of methods indicates that the lake in question (where the fish were caught in the summer of 1969) had been contaminated by *both strontium and cesium.* Since the article was published in the section of the journal called "Short Communications," a detailed description of methods was not to be expected. However, even this brief report gives the following quantitative data:

> The concentration of strontium-90 in the water of Lake X was 0.2 microcuries per liter.
> The concentration of cesium-137 in the water was 0.025 microcuries per liter.
> Only four species of fish inhabited the lake:
>
> roach (*Rutilus rutilus L.*.)
> ide (a fish of the carp family, *Leuciscus idus L.*)
> perch (*Perca fluviatilis L.*)
> pike (*Esox lucius L.*)

For their study of radioisotope concentrations, the researchers caught only roach and pike (pike feed on roach). Cesium was found mainly in the muscles, and strontium in the bones. A total of forty-four roach and thirty-two pike were taken during the summer of 1969 and analyzed for cesium-137 and strontium-90 content.

We must first point out that the concentration of strontium, ten times higher than the concentration of cesium, was also *higher than the maximum permissible doses of strontium in water to be used for drinking or for commercial fishing according to Soviet standards* (14). Safety standards found in all basic guidebooks in the USSR and in the Soviet Medical Encyclopedia (1968) state that the highest permissible concentration of strontium-90 in lakes open to the public or in sources of water supply cannot exceed 3.10^{-11} curies per liter. This level is approximately 5,000 times lower than the level in the lake under investigation. In experimental bodies of water, this dosage could unquestionably be increased, but 0.2 microcuries per liter is much higher than necessary for merely experimental purposes.

From Ilenko's survey and monograph (13,14), it is evident that as early as 1960–61 he did experimental studies on the distribution of strontium-90 and cesium-137 in fish inhabitating a lake contaminated by these isotopes. In this case the author had half the level of strontium but four times as high a concentration of cesium. This was an experimental body of water into which radioactive phosphorus had also been introduced. In tests made by other authors for experimental purposes much smaller concentrations have been used.

However, in Ilenko's 1970 paper there are no data enabling us to judge the size of the lake. The fact that thirty-two pike and forty-four roach were caught is not a sufficiently reliable indicator. In the Urals (Chelyabinsk, Sverdlovsk, and other areas), there are five-thousand square kilometers of lakes and ponds which are used for commercial fishing, and the average productivity of the Urals lakes is 16–25 kilograms of fish per hectare. However, pike usually constitute from 2 to 6 percent of the catch (15). The pike in Ilenko's experiments were from 3 to 6 kilograms in size; roach are not sizable at all. By rough estimates, the researchers caught about 150 kilograms of pike during the summer. For ecological reasons no more than the commercially allowed number of fish could have been caught. Therefore, we may assume on a preliminary basis that the lake referred to in the article was no less than a hundred hectares in size, but its actual dimensions could have been much greater.

After two years, Ilenko published a much more detailed paper (11) on the basis of research done in the same area and on the *same lake*. There is absolutely no question about this, since the 1970 paper is referred to as a "preliminary report" on the same research. However, in the more detailed work only the distribution of cesium-137 through the food chain is studied, not that of strontium-90. A later book by Ilenko makes it clear that no further studies of strontium were done in that lake, only of cesium. The reason for this selectivity is not made clear, nor is it scientifically justified since both isotopes have a different localization and a different type of exchange in the composition of plants and of fish. It is also strange that neither in the 1972 article nor in the book does Ilenko mention that there was radioactive strontium in the water as well as cesium. The entire presentation is made as though cesium-137 were the only radioisotope in the lake.

The article has no section on methodology, and both the description of the lake and the explanation of how the radioactive cesium came to be in it *are obviously falsified, although apparently the author was obliged to do this*. Unfortunately we shall encounter such deliberate falsification many times. The fish (pike and roach) were caught over a period of two years beginning in June 1969 and ending in December 1970. The preliminary report in 1970 deals only with the catch in the summer of 1969. In his introduction (11, p. 174), Ilenko indicates that it was a non-running-water lake and that the concentration of cesium varied seasonally as a result of *artificial* changes. It is unclear how this is to be interpreted, but we are apparently meant to assume that cesium-137 was added precisely in 1969 and 1970. (Nothing is said about variations in cesium concentration in the 1970 report.) The variations of cesium in the water and in the bodies of the roach and pike are presented by Ilenko in the form of a graph, which we reproduce here (Figure

2). It can be seen that the concentration of cesium-137 in the lake changed almost every month. From June to September 1969 the concentration of cesium rose steadily from 0.005 microcuries per liter to 0.02 (a fourfold increase). If we are to believe the "introduction," this would have to do with the constant addition of cesium to the lake. However, in the autumn the concentration continued to mount, reaching its highest point in December (0.04 microcuries per liter). It is incomprehensible why it was necessary to increase the isotope content in the winter, when lakes freeze over and the biological activity of fish virtually ceases. By the spring (April 1970), the cesium concentration in the lake had dropped sharply (an eightfold decrease), but in May it increased, then declined a little in July 1970, and doubled once again in August (reaching 0.04 microcuries per liter). By December, however, it was back down below 0.01. These sharp variations, if they were experimental, would have no logic, and for a "non-running-water lake" they are quite impossible. The lake, as we shall see, was fairly large (several square kilometers). In this kind of lake the isotope concentration could increase if introduced from an external source, but according to Rovinsky's formulas (9), the sudden sharp reduction of 800 percent over a two-to-three month period in the spring of 1970 would have been impossible in a *non-running-water lake*. If we grant that this could have happened through the action of the biomass (which is doubtful in the case of the especially sharp decline from December to March, when the lake was covered with ice and the water temperature was 3–4°C), the reduction by December 1970 could not be attributed as well to an increase in biological absorption. The only way cesium could disappear from a lake in such quantities would be from the *inflow of fresh water* typical of a *running-water* lake. Unquestionably, the cesium contamination (and the strontium as well) came from a *powerful* external source, whose activity was not regular, but apparently depended on climatic factors (precipitation, the flow of ground water, etc.). The process of inflow of fresh water existed, that is, it was a running-water lake, *which Ilenko could not reveal*. If the lake waters ran off (and this relates to the strontium-90 as well), the question arises, *Where to*? The runoff from lakes in the northern part of the Chelyabinsk region (the Kyshtym district) and from the Sverdlovsk region passes through a system of small rivers and enters the mighty Ob, which flows into the Arctic Sea. The run-off from the southern part of the Chelyabinsk region and other parts of the southern Urals goes into the Caspian Sea. In both cases for a distance of thousands of miles along the route of the runoff there would be biological and chemical fixing of strontium and cesium. To report this type of radioactive contamination of major rivers would be impermissible for reasons of censorship.

It is difficult to answer the other question, Where did the strontium go? Ilenko did his first paper quickly in the summer of 1969. Water samples were taken and its activity measured (0.2 microcuries per liter for strontium-90 and 0.025 for cesium-137). Then for two or three months fish were caught and the levels of strontium in the bones and cesium in the muscles were measured. The results were processed quickly in Moscow. (Ilenko works at the Institute of Evolutionary Morphology and Ecology of Animals under the Soviet Academy of Sciences in Moscow.) Then they were sent to a journal and published by mid-1970. From September to December of 1970 no fish were caught, and in December only pike were caught (very few apparently, because of the difficult conditions of ice fishing).

The work of the group continued in December 1970. (To judge from the number of measurements and the scale of the operations, a large group of technicians was required for this work and unquestionably Ilenko only supervised it.) The group found a sudden increase in the concentration of cesium (and of strontium as well of course), but the taking of fish had not been systematic and the analysis of the biomass had only begun in the summer of 1970, when the first article had already been published. In the spring a sharp decline in the concentration of isotopes began (the inflow always increases in the spring because of melting snow). But the lake had already been called a "non-running-water lake," and Rovinsky had demonstrated a different dynamic over time for *strontium* in non-running-water lakes. From the cesium alone, it was obvious that this was a running-water lake, and the strontium dynamics could not coincide with the cesium dynamics—the different behavior of these isotopes had been well known for a long time. That would have exposed the falsity of the assertion that the contamination was "artificial." If curves for the two isotopes were given in the same work, it would show clearly that there could not have been any "periodic, artificially changed cesium content." The strontium had to be sacrificed to ensure the possibility of publication. Together these isotopes would have complicated the picture too much. Moreover, their patterns of distribution within aqueous systems are quite different and had already been studied well enough in model experiments. The strontium content in such a lake would gradually be reduced over a one- or two-year period.

The author does not give any data on the size of the lake, but it is possible to draw some conclusions indirectly on the basis of the number of pike caught. These are predators and their number is limited by the presence of the fish on which they feed. In the lake being studied the pike fed *exclusively* on roach; there were only four species of fish in the lake, three of them predators.

The lakes of the Urals and Siberia most often contain a far richer selection of fish species (15), but there are also "poor" lakes, in which precisely perch, pike, and roach predominate. Ilenko studied food chains between the plant and animal worlds of the lake and within the "fish" biocenosis for a two-year period. In such research the number of pike taken should not disturb the normal balance between predators and their food sources, that is, the total stock of pike in the lake should not be substantially changed. During the two years of observations,

Ilenko caught more than one hundred pike, most weighing between three and five kg. and several weighing between ten and twelve. The presence of such large pike in the catch is evidence that the lake had not been used for commercial fishing for many years, which allowed some of the pike to grow to that large size. The total weight of the fish caught was about four to five hundred kg. The symposium volume *Biological Productivity of Siberian Lakes* (1969), from which we have already cited an article about the Urals (15), gives the composition of fish in lakes of various types. Lakes in which only roach, perch, and pike predominate actually are the main type in the Sverdlovsk and northern Chelyabinsk areas. Lakes located farther south have a richer vegetation and a more complex fish population, with sometimes as many as thirty-six different species (including bream, carp, pike-perch, and various kinds of whitefish). The productivity of lakes with a rich fish population is higher than that of poor lakes. The productivity also depends on the depth of the lake, which in this case we do not know. In many lakes the productivity is 10 kilograms per hectare; in running-water lakes it is between 16 and 25 kilograms per hectare. In poor lakes where only three or four species live, such as the lake in Ilenko's experiment, roach constitute between 80 and 83 percent of the catch (and of the species balance in the lake as a whole), perch constitute 3–10 percent, and pike 2–6 percent. If we take an average figure of 4 percent for pike, we would get only one kilogram per hectare in an ordinary "poor" lake. But in 1970 Ilenko caught 300 kilograms of pike, which indicates a lake no less than three hundred hectares in size, that is, no less than three square kilometers.

But in population research and in food-chain analysis, a more cautious method must be followed than in commercial exploitation of lakes. There had to be absolute certainty that the taking of one hundred pike would not disturb the ecological balance of the lake (roach were taken in smaller quantities in relation to their supply in the lake). An ecologist works without changing the biological equilibrium, and therefore Ilenko had to be sure that one hundred pike would not exceed 5–10 percent of the population. But there would be a stock of between a thousand and two thousand pike in a lake of this "poor" type only if its size was approximately ten square kilometers. Thus, as in the case of Rovinsky's "experimental" lakes, Ilenko's research was done on a lake of *geographical* size, not on a small pond. And the above calculations refer only to the minimal possible size.

The Quantities of Radioisotopes in the Lake

In a lake of this size and on the assumption that the lake was shallow, as most Urals lakes are, about 5,000 curies would have been needed to increase the cesium-137 level from 0.01 to 0.04 microcuries per liter from September to December 1969. For the two other increases in 1970 about 5,000 more curies must have been added. But the concentration of strontium-90 was even higher (0.2 microcuries). Consequently there were no less than 20,000 curies of strontium-90 in the lake water. This is unquestionably not an experimental quantity; it is on an industrial scale. But this is only the water. The bottom deposits and the biomass (algae and plankton) must have accumulated much larger quantities of radioisotopes, thus providing for the "biological cleansing of the water." Figure 2 shows the pattern of concentration of cesium-137 in the food of the roach (plankton and algae). There are quite large variations, indicating change in the water of the lake. But in August 1970 the concentration in the food of the fish was 38 microcuries per kilogram. According to Ilenko's calculations, "The absolute quantity of celsium-137 in the food of the fish substantially exceeds the concentration of the isotope in the water (by a factor of 520–4,200, with an average factor of 1,300)" (11, p. 175). There are no figures for strontium in the food. Nevertheless, to judge from the data in Rovinsky's paper, the quantity of the isotope in the bottom deposits and silt (the biomass was not counted) was dozens of times higher (in absolute figures) than in the water.

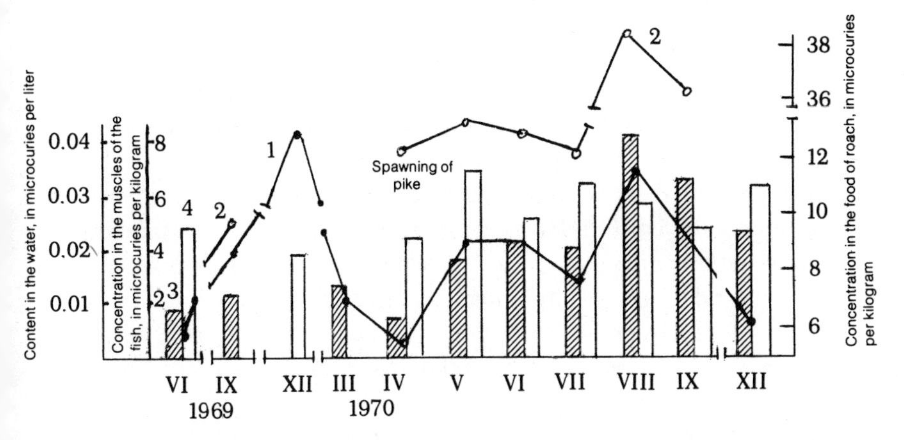

Figure 2. Cesium-137 content in the water of a lake (1) and its concentration in the food of roach (2) and in the muscles of roach (3) and pike (4). (Ilenko, 1972 [11].)

The cesium level in the muscles of the fish was 5 to 10 microcuries per kilogram and the strontium level in the bones was five to ten times higher, which unquestionably made the fish of this lake unsuitable for human consumption.

If the figures indicated by Ilenko for the absolute concentrations of radioactivity in the food of the fish are fairly reliable, and if we consider the variations (three abrupt increases in two years), we arrive at the fantastically high figure of 10 million curies for cesium-137 alone. This is the equivalent of 10 tons of radium. And of course there was also strontium in substantially larger quantities. Such quantities (measured in megacuries) are found only in reactors. There is no real possibility that Ilenko, even with a group of unnamed associates, could carry out biological experiments with such quantities of radioactivity. In all these calculations, the silt deposits have not been taken into account, and these, according to Rovinsky's data, fix an even larger amount of radioactivity.

Of course our calculations are highly approximate, since there are no data for the variations in the strontium level. In 1969 Ilenko indicated a strontium level ten times higher than that for cesium. If it was a running-water lake (and that follows from the dynamics of the isotopes), the estimated amount of radioactivity being discharged into the lake should be increased. The year in which the discharge of radioactivity into the lake began is not made clear, but it is evident that there was radioactivity before 1969. In June 1969, when Ilenko made his measurements of cesium, its concentration in the pike was approximately the same as in June 1970. It is somewhat paradoxical that the pike muscles in almost all measurements (except two) contained a larger concentration of cesium than the muscles of the roach which serve as the pike's food. The concentration of cesium in the muscles of the large pike (10 to 12 kilograms) was two times greater than the concentration of cesium in the pike weighing 3 to 5 kilograms. The half-life of cesium in mammal organisms is 150 days (16). In fish the exchange takes place even more slowly, especially if we allow for the winter rest period. The high level of cesium in the muscles of pike, when the measurement began in June 1969, tells us that the accumulation began long before 1969 and that apparently in the preceding years the concentration of isotopes in the components of the lake was even higher, as reflected in the final link of the food chain (pike) and to a greater degree in the larger (older) pike.

All these data (the exact size of the lake, the total quantity of radioisotopes introduced into it, the location of the lake—necessary to calculate climatic factors—and many other details) should have been indicated by the author as a matter of course. Without these, the entire work loses its ecological value. All the conclusions reported in the article were essentially known from previous experiments in small bodies of water under artificial conditions. It is precisely the uniqueness of the conditions and the large size of the lake that has value in Ilenko's work, but the author makes a conscious effort to conceal this *uniqueness,* and sometimes even falsifies the description of the actual conditions of the experiment.

If one thinks through the implications of these papers, it becomes quite clear that the contamination of the lake either was the result of a serious accident or it involved the discharge of industrial reactor waste into the lake by a large local nuclear plant, such as a reprocessing plant. The latter explanation is highly unlikely for the years 1969–1970, however, since the regulations for the discharge of radioactivity into the external environment had become very strict by that time, and the discharge of millions of curies of strontium and cesium into a nearby lake simply would not have been permitted. A more likely hypothesis is that the lake and the territory around the lake were contaminated as a result of an accident, resulting in the natural discharge of radioactivity from the surface runoff and ground water. If this is the source of the contamination, it explains the variation of cesium in the water—in the spring of 1970 the melting snows sharply reduced the relative radioactivity of the water, and from April to August the radioactivity level increased with the inflow of more radioactive ground water. The fall of 1969 could have been dry and the fall of 1970 might have had an abundance of rain and therefore the surface runoff into rivers and lakes could have reduced the level of radioactivity.

In confirmation of the hypothesis that the radioactivity in the lake and its variations were linked directly with the contamination of the surrounding territory (the lake's drainage basin), there is one indisputable fact. Ilenko and his group did extensive research during the same time (from the summer and fall of 1969 and 1970) on the distribution of the same radioisotopes, strontium and cesium, in land animals and birds, trapping and shooting them by the hundreds. Since the two research programs were carried out simultaneously by the same group, and since the fish and animals were taken continuously, the territory used must certainly have been adjacent to the lake.

Chapter 5

Mammals in the Radioactive Contaminated Zone of the Urals

It would be logical to take up the question of surface and soil animals living for many years in the radioactive biocenosis by starting with the lowest forms—soil worms, ants, snails, insects, etc.—before moving to amphibians, reptiles, birds, and mammals. Even before that, a general review of the radioecology of plants would probably be in order. But I have begun with fish and am now going on to mammals because the present work is not simply a review of radiation botany or zoology but an analysis of a particular event. For that reason, we should first examine the research and facts that best reveal the nature of the nuclear disaster. After that the data on insects or soil algae will be more comprehensible, and it will become evident why such an unusual selection of doses of radioactivity was used in this research and why the methods diverged so substantially from the normal *experimental* methodology, which certainly would have been followed if the research had actually been planned in advance.

The radioactive biocenosis in the Urals is apparently the most extensive in the world, but by no means the only one. Contamination of large areas by radioisotopes has occurred in other countries too, partly as a result of unfortunate side effects from atomic bomb tests (for example, the release of radioactivity from an underground test in Nevada, about which I will speak separately). Another example was the appearance of a 16–17 hectare area contaminated by radioactivity near the Oak Ridge National Laboratory in Tennessee, at the site of a lake which had been drained between 1956 and 1958 after radioactive waste had been discharged into it. And there have been other examples. These contaminated areas came to serve as experimental bases for dozens of radioecological studies in which the effects of isotope absorption on plants, animals, and food chains was traced. These studies dealt with both higher and lower animals, plants, and microorganisms, and were published in ordinary journals with highly detailed descriptions of methods, the sources of the contamination, its history, and many other experimental details, about which almost nothing is said in the studies by Soviet authors carried out in the southern Urals region.

In preparing their works for publication, Soviet authors—for example, in the fields of biochemistry or physiology—have long followed established international standards in describing the methods and conditions of their research, to ensure the possibility of its replication. The absence of such "openness" (and sometimes its replacement by obvious and deliberate pseudo-information) is plainly evident in the works of the research groups which studied the Urals contamination. In view of this, before beginning an analysis of Soviet papers dealing with mammals, I will refer the reader to several studies done in the United States on the very same topics, but somewhat earlier (17–19). These may serve for methodological comparison. I have selected these works, on the results of the contamination in Nevada and elsewhere, at random from dozens of others simply to show that research in this field was begun long ago and that the first publications on mammalian food chains in radioactive environmental conditions date from the early 1960s. It is possible that this in fact was one of the strongest incentives prompting the Soviet authors, who had worked in a substantially more extensive and varied radioactive ecosystem, to begin publication of their own data. After all, researchers in secret laboratories have the same scientific ambitions as others, aspiring to be first in making some discovery known to the world.

The first work I will single out from the general flood of Soviet scientific literature on radioecology was published in 1967 by A. I. Ilenko, whom we already know from the previous chapter, and his colleague G. N. Romanov (20). The author presents data on seasonal and age changes in the accumulation of strontium-90 for *only one* species of field mouse (*tyomnaya polyovka*), which lived under natural conditions in areas contaminated by strontium-90 in doses from *1.8 to 3.4 millicuries* per square meter. I have especially stressed the level of contamination (from 1.8 to 3.4 millicuries per square meter) because, later on, this radiation level will help us recognize many other studies carried out by other authors and with other research subjects but in the same radioactive ecosystem. Other contamination levels, too, will be encountered repeatedly in several different groups of related research papers (with the doses always in the chaotic and random pattern typical of this kind of industrial contamination).

Ilenko and Romanov do not indicate *exactly where* the research took place, and the research subject they have chosen is found virtually throughout the territory of the USSR. It is important, however, to note the authors' statement that their observations were carried out during *1964 and 1965*. In the contaminated zones, the mortality of the mice was higher than in the control areas. Mice born at the beginning of the summer were weaker by autumn and had a higher mortality rate during the winter than mice born in the same territory at the end of the summer.

In another article (21), Ilenko included a schematic diagram of the fenced-off enclosure in which mice were taken for study. It shows a scattering of various levels of activity (from 1.8 to 3.4 millicuries per square meter) in the contamination of the soil. The same diagram is reproduced in his book (12, p. 37). I reproduce this diagram here in Figure 3 because the random and capriciously mosaic distribution of zones with different levels of activity obviously attests to the fact that the radioactivity was introduced in this territory accidentally and not according to any

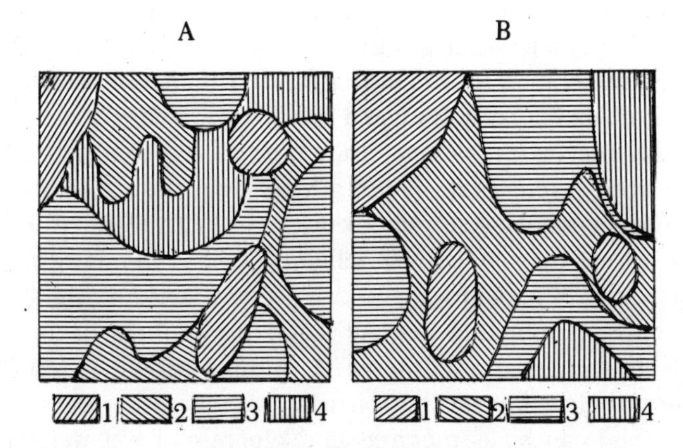

Figure 3. A. Average strontium-90 content (in microcuries per gram of dry weight) in plants in an enclosure; B. strontium-90 contamination of the soil in the enclosure (in millicuries per square meter).

experimental design. The size of the enclosure was one hectare, but the date on which the contamination occurred is not indicated (only the fact that the mice had already lived in this radioactive environment for a long time). In another diagram, Ilenko gives a botanical description of the area: a young birch forest under whose canopy were five different types of grass cover. Fifty mousetraps were set around the area. The distribution of strontium-90 in the plants and mice corresponded to the levels of concentration in the soil, which is not surprising and could have been predicted.

One methodological circumstance is rather strange—the absence of data on the radioactive contamination of the soil not in terms of square meters but of soil depth. Strontium-90, if applied superficially, fixes firmly in the topmost layers of the soil. Therefore both the way it is applied and the *depth* of the contamination are important ecological factors. A slight increase in the depth of the application reduces the surface radiation and changes the pattern of strontium uptake by plants. Absorption by trees is increased if the contamination penetrates to a greater depth, but at a lesser depth there is greater absorption by grassy plants. In addition, the figures 1.8 to 3.4 millicuries per square meter of soil for 1964–65 could not be typical for 1962 or 1963. For that reason, the dosage of the *initial contamination* should have been given. From nothing more than the data about absorption by plants—0.25 microcuries per gram of dry weight in areas where the soil had 3.4 millicuries and 0.10 microcuries in areas with 1.8 millicuries—we can see how arbitrary the figures for soil contamination are. Although the authors do not give absolute data on the volume of plant mass per square meter, it is nevertheless known that plant mass (when there is nothing but grass) can be as much as one kilogram of wet weight and 100 grams of dry weight per square meter. But the trees, too, draw strontium-90 out of the soil. The dry weight of trees, even saplings, is measured in kilograms per square meter. Thus in the plants, taking the total for both trees and grass, we get an even larger amount of radioactivity than in the soil (about 3 to 10 millicuries per square meter). *From this it is evident that the figures for soil contamination were ascertained for only a certain upper layer and not for the entire soil mass in these areas.*

In these first studies, the field of one hectare in which the observations were made was fenced off from the rest of the area, so that changes in the strontium accumulation in the mice could be measured by cutting off bits of the mice's tails periodically. Each mouse (of a total of about three hundred) was under observation for two years, and it was possible to follow their mortality rate and the relation between that and the accumulation of strontium-90 in the skeleton. The experimental plan required an *isolated* area, but if the introduction of the isotope into this limited area was truly experimental, any experimenter, following the general procedures for field tests, would have broken the territory down into a series of isolated quadrants of equal size, each with a different level of activity (for example, 0.1; 1.0; 2.0; 3.0; and 4.0 millicuries). In a *planned* study it would make no sense to set up contaminated zones in a random pattern with relatively close upper and lower radioactivity limits (1.8–3.4 millicuries). Such a distribution of isotopes could only be explained by the fact that the territory was already contaminated before the research was begun.

A year later, Ilenko published a new paper (22) in which strontium accumulation in nine different species of small mammals was measured. The aim of the experiment was to compare strontium-90 levels in the skeletons of various species. In this case the author was not required to make observations several times on the same animal. Animals caught in traps were killed and the strontium-90 in their hipbones was determined. Thus the observation sites were not fenced-off. We are not informed how these measurements were made, but since the article was submitted to the journal in mid-1967, it can be assumed that the work was done right after the previous study, that is, during 1965 and 1966. The patterns of radioactivity distribution in these sites are not given, but they were totally different from the previous sites, since the animals were caught in zones with contamination densities of 0.6; 1.0; and 2.5 millicuries per square meter. The size of the total area contaminated by strontium-90 is not indicated, but the area in which the animals were caught certainly had to be larger than one hectare, since 1,066 specimens of nine different species were killed for measurements. Since the connection between the strontium level in the bones and the type of food eaten by the animal was being determined, the taking of 1,066 specimens should not essentially disturb the population balance in the given environment. If approximately 10 percent of the animals of the biocenosis were taken, the *minimal* dimensions of the contaminated territory, in view of the total number of animals of all species upon which observations were made, had to be in the range of one to two hundred hectares. The nature of the vegetation in this territory and its locality are not specified.

Two conclusions to be drawn from this research should be kept in mind. First, the strontium contamination seems to have been only in the surface layer of the soil, because the smallest accumulation was in the skeleton of the common hamster, which feeds on seeds and the underground parts of plants. Second, there were "clean" biocenoses not far from the territory in question. This follows from the fact that among the *migratory* species, there occurred specimens with hardly any strontium-90 in their skeletons. The author relates this to their possible recent arrival from "clean" territory. It is also undoubtedly true that the reverse migration of "dirty" (contaminated) animals into the "clean" zones was possible.

The territory on which these observations were made apparently had a soil composition different from that in the previous study, since the concentration of strontium-90 in plants was ten times higher. (The loss of strontium-90 varies greatly with the type of soil.) The zones with different contamination levels were not all of the same size and undoubtedly were located in differing ecosystems, because the species mix of the animals varied from zone to zone. The largest was the one contaminated by the highest concentrations (2.5 millicuries per square meter). Eight species were taken in this sector, among them 108 red-toothed shrews of the species *Sorex caecutiens* and 40 of the species *Sorex araneus*. These animals feed mainly on earthworms in the soil and migrate over great distances. *Sorex caecutiens* did not appear in the traps in other sectors. The second zone was apparently in the forest since among the four species taken there, woodmice (*Apodemus sylvaticus*) predominated, 308 of them being taken, by comparison with 30 in the zone of highest contamination and 45 in the zone with slight contamination (the data for strontium concentration in plants were given only for the zone with the highest contamination).

All this indicates once again that the contamination was accidental, not planned in advance. Moreover, the over-all contamination of such a territory (judging by the strontium in the plants) would require approximately 4,000 curries (Ci) of strontium-90 at the minimum. In planned experimental work, the same findings could be obtained by using two to three hectares, or an even smaller area. It would not be a rational act to contaminate one to two hundred hectares of unfenced territory for hundreds of years, especially since migratory mammal, bird, and reptile species, as well as seeds and pollen, would cause a secondary dispersion of strontium over much greater distances.

In 1969 Sokolov and Ilenko (23) published a survey article on vertebrate radioecology, giving previously unpublished experimental data, not only for strontium-90 *but also for cesium-137*. But they did not indicate the methods used in their experiments, or cite previously published works in which such methodology could be found. (In his monograph [12] Ilenko again presents the same data but refers only to the survey article.)

In reference to strontium-90, Sokolov and Ilenko cite data (23, p. 249, Figure 4), in addition to the tables given earlier in experimental articles, on the population size of four different species of mice in plots with varying contamination levels (1 to 3 millicuries per square meter). Once again, this is a new territory because the type of vegetation differs from that in the areas previously studied (again, the sizes of the new plots are not specified). Three types of plant cover were examined: the edge of a forest (3 millicuries per square meter), thickets of weeds (2 millicuries per square meter), and meadows with bushes (1 millicurie per square meter). Again, it is quite obvious that the area used for these experiments had been contaminated earlier, because in studying the population density in relation to the contamination level (which was the research aim) the researchers would *necessarily* have to test different contamination levels in each ecosystem, that is, to have nine plots altogether, three in each system, not just three different areas. A territory contaminated by cesium-137 appears for the first time in this article, and findings are made for the content of this isotope in six different species of mice in plots with three different levels of cesium-137. No methodological details are given, but it is quite obvious that all three fields had differing vegetation and ecologies, since they were not homogeneous in their species mix. Only two species could be caught in the first field; four were caught in the second and third, but the two additional species in each case were not the same. The level of cesium-137 contamination in these fields was no longer measured in millicuries but in *microcuries* (7.85, 5.30, and 4.45 microcuries per square meter), that is, there was approximately five hundred times less cesium-137 contamination than the maximum for strontium-90. Whether this was the same territory where the strontium was found is not clear. It is quite easy to obtain different readings for the radioactivity of strontium and cesium, since the two isotopes have different types of radioactivity (beta radiation as opposed to gamma radiation). The fact that the cesium contamination was not experimental can be guessed from the closeness of the contamination levels in the different plots. If these were planned experiments, more pronounced differences in the contamination conditions would certainly have been introduced.

In the first stages of fission in nuclear reactors (and in atomic explosions), there is not such a substantial difference in the amount of strontium and cesium produced. Therefore if a territory is contaminated by accidental local fallout from weapons tests or by *fresh* reactor waste from an industrial accident, the strontium-90 and cesium-137 content in the soil would not differ by a factor of 300–500. Cesium is an analogue of potassium and this isotope does not become so permanently fixed in the biomass (or, apparently, the soil) as strontium-90. Therefore, the cesium-strontium ratio could be expected to decline over time. But not very quickly. However, in the processing of waste before it is stored cesium-137 is often extracted, because its gamma radiation may find practical radiological application. Cesium radiation sources, because they have longer half-lives, are more convenient than cobalt sources. And in the case of a gamma-emitter used for medical purposes, a hospital would need thousands of curies. It is also possible that cesium gamma radiation has practical uses in fields other than medicine. If there are processes for extracting cesium as well as plutonium and uranium at nuclear industrial centers, the stored waste would of course contain much more strontium than cesium. The fact that the strontium-cesium ratio varied from 10:1 to 300:1, both in Lake X and in the soil, can only be interpreted to mean that the contamination involved the release of wastes from different phases in the processing of the by-products of the atomic industry.

It is quite likely that *because of* the very low levels of cesium in the biocenoses being studied (levels ranging from 4 to 8 microcuries per square meter), the *radiobiological* and genetic effect of the exposure of plants and animals to cesium was too minor and was often simply ignored. Despite the beta radiation of strontium, the presence of that isotope in concentrations one to three hundred times greater than the cesium concentrations and the fact that strontium fixes more firmly in biological structures, made that radioisotope the dominant radiobiological factor, clearly affecting the physiology, genetics, and population ecology of both plants and animals. This is confirmed by the data of Ilenko (12), who studied numerous morphological and physiological changes in animals inhabiting areas with strontium levels of 1–3 millicuries per square meter. Because of these serious effects, such high doses should not have been used to study *food chains*. But apparently the authors did not have a choice.

If we figure in the data on cesium-137 and new data relating to the dependence of population density on the strontium level and the type of biocenosis, the total number of different kinds of mice caught for this research increases to fifteen hundred. Here, too, we must assume that only part of the population was caught, so that the normal food chain balance was not disturbed. (Provisionally, I estimate this to be 10 percent of the total population, although it may be even less.) According to the data of L. Nikitina (24), who studied migration patterns for various species of mice and other rodents in different parts of the USSR, the "population density" of the main species in the Urals forests—the red field mouse (*Clethrionomys rutilus Pall*)—varies between twenty and eighty animals per hectare. Ilenko set as many as forty-eight traps per hectare. Standard types of traps are used to study mice in the forests and other parts of the Urals, and there is a statistical average for the number of mice ordinarily caught: one to two mice per one hundred traps per day (25). These data allow us to place the probable dimensions of the contaminated territory on the order of hundreds of hectares, and that is only the *minimum* estimate. Thus, by simply examining the studies that were done one after the other, we find that the size of the contaminated zone systematically increases to dimensions involving square kilometers (100 hectares = 1 square kilometer). And as we shall see, this is by no means the outside limit of the contaminated zone.

Ilenko published a separate article at a later time, on cesium-137, coauthored by E. A. Fedorov (26). Once again, there was no methods section, but the authors acknowledged in their introduction that cesium was not the only radioactive isotope in the contaminated territory. "The research was done in experimental plots contaminated by radioactive substances (modeled after an industrial contamination), *one of which* was cesium-137" (26, p. 1371). (Emphasis added.)

The cesium-contamination level in these plots was 4–8 microcuries per square meter. The contamination level of the other isotopes, however, was not specified. The purpose of the paper was to study food chains and the concentration of cesium-137 in the bodies of *twenty-two* animal species. The work was conducted on a scale larger than in all the previous research. In addition to small mammals, large ones were also shot, for example, rare fur-bearing animals—the Siberian ferret (*Mustela sibiricus*) and ermine (*Mustela erminea*)—and roe deer (*Capreolus capreolus*), as well as several types of birds. From the species mix it is easy to see that the work was done in Western Siberia or in the Urals, because some species are not found in the European part of the Soviet Union. Ermine are so rare, and their fur, so unique, that in olden times it was used to fringe the robes of the tsars. A large territory would clearly be required to obtain specimens of this rare animal, without disturbing the population balance. However, the chief index of the size of the territory was the deer. Observations went on for two years, and there had to be a certainty that all the animals remained within the limits of the radioactive territory. The fact that five deer were shot indicates that the territory contained a deer herd of at least thirty to forty. The feeding area for one deer in the summer is no less than forty to eighty hectares; in the winter, when thick snow cover makes feeding more difficult and tree bark is the main food, deer may migrate many kilometers. Thus, it is evident that the total territory contaminated by cesium must have covered *thousands of hectares* at least.

In the same year (1970), Ilenko published a more detailed article (27) which made it quite clear that the areas studied in 1967–69 for strontium distribution and the regions referred to in other articles (23, 26) as contaminated by cesium *constituted a common territory*.

The author measured the concentration of both strontium-90 and cesium-137 in the animals simultaneously; the figures for such concentration in food, in skeletons, and in the muscles given in the tables show that the data published earlier were being repeated.

For strontium-90, the contamination was between 0.6 and 2.5

millicuries per square meter. This is the same as in an earlier study (22), but without any division of the area into subsectors. It was possible to divide the territory into small experimental plots with varying activity of 0.6, 1.0, etc., for small rodents (mice of various species) and for plants. Mice do not migrate far from their nests, and of course plants are stationary. But in determining the radioactivity for roe deer, reindeer, and other mammals that migrate over great distances, or for birds, there is no feasible way of applying doses in such a differentiated pattern. This indicates for the umpteenth time that the work was not planned in advance, because under experimental conditions an uneven distribution of radioactivity over a territory would place impediments in the way of studying *food chains*. The problem would be especially serious in relation to predators, because predator families usually have a particular "hunting" territory, and a particular contamination level should be applied to coincide with such a territory. This obviously was not done.

The author unquestionably understood the methodological difficulties: in his paper he gives the figures for strontium concentration in the skeletons of the various species only in an averaged-out, "summary" form. In the work with mice, as we have seen, the radioactivity was determined for each individual.

For cesium, Ilenko indicates a contamination level of 4.6 microcuries per square meter, rather than "4–8," as in the other publication (26), but the figures for plants and animals are often taken from tables published by Ilenko and Fedorov (26), though interpreted by a quite different method. For example, the concentration of cesium-137 in the dark field mouse species in both studies is 2.8 microcuries per kilogram in the body, 5.4 in the food, and in the case of roe deer, 0.4 in the body. The same figure was given for the number of deer shot in order to measure cesium levels—that is, five. For other animals the author averages out the earlier data, which were broken down according to group; nevertheless, in the case of birds (magpies, starlings, field sparrows, and others) the figures for the number shot and for the cesium concentrations in them are identical. To determine the strontium levels, eleven deer were shot.

In the introduction, the author states clearly that the territory was contaminated by *strontium and cesium.* To judge by the figures, the cesium and strontium measurements were made separately and independently; sometimes most of the animals were killed to make cesium readings, sometimes for strontium readings. Therefore it is not excluded that the total number of deer shot for analysis was sixteen. There were identical statistics in the two works for cesium in frogs and in such fur-bearing animals as the ermine and Siberian ferret.

The comparison of these data, then, make clear that the territory was large enough to support several thousand animals of different species, including a large herd of roe deer (sixteen deer having been shot apparently without disturbing the natural balance of the population). The territory was contaminated by strontium at levels from 0.6 to 2.5 millicuries per square meter and by cesium at levels five to six hundred times lower. Going by the crucial species (deer), we can assume a territory between fifty and one hundred square kilometers in size, one that can no longer be measured in hectares. To reach the level of strontium indicated in these works, some 500,000 curies must have been spread over the territory. If the strontium level in plants is included, this figure rises to *one million curies*—unquestionably an industrial, not an experimental magnitude. No experimenter in any country would contaminate dozens of square kilometers with strontium at levels far above those used in tracer doses—levels that would produce many morphological changes, increased mortality rates in animals, and a number of other serious consequences.

Usually the authors say nothing about the location of the experiments or the size of the total area studied. But if we compare the information in the various publications, the methodological falsification becomes quite plain. For example, Ilenko repeats the same data in three of his studies (22, 26, 27) published in 1968–70 but does so in different contexts and gives different explanations for the contamination of the territory each time, sometimes contradicting himself. One of Ilenko's works (27) indicates that the territory was contaminated by both cesium and strontium. Why there was such a big difference between the strontium and cesium doses used in the "experiments" is left unexplained. Nor is there an explanation of how the contamination occurred. At the same time, the work by Ilenko and Fedorov in which measurements were made for cesium alone (26) states that the contaminated areas "were established to work out methods for the dosimetric monitoring of objects in the external environment (Korsakov et al., 1969)."

Thus the authors assert that contamination was introduced by a different group of researchers engaged in dosimetry. However, a look at the work by Korsakov et al. raises strong doubts that Ilenko and Fedorov actually read that work. They seem to have picked up an accidental reference to it or to have heard about it by some other indirect means. For they refer to it *as though it were a foreign publication* and give the title in English. But the article by Korsakov et al. (28) was published *in Russian.* International agencies connected with the United Nations recognize Russian as a working language and publish Russian papers without translating them—and the IAEA in Vienna is such an agency. The authors of the article work for the Kurchatov Institute of Atomic Energy in Moscow. If their descriptions are to be believed, their work was genuinely experimental and consisted in the contamination of a territory with tracer doses using a mixture of fission products "which had been stored for 200–350 days after being irradiated in a reactor." (Why if there was only *one* reactor, do they speak of 200–350 days, and not a more precise period?) In other words, this was reactor waste after almost a year of storage. The radioisotopes used included cesium-137 and strontium-90, and such others as cesium-144, zirconium-95, and ruthenium-106.

These radioactive substances were scattered over a fairly

large territory (including populated villages and agriculturally productive areas), but the contamination level was one curie per square kilometer. *In that case there would be one microcurie of the isotope mixture per square meter.* In what part of the USSR this work was carried out is not specified. But in the work by Ilenko and Fedorov the cesium activity alone was 4–8 times greater, and for strontium it was 1,000–2,500 times greater, than in the report presented by Korsakov et al. to an international symposium. Since detailed analyses of radioactivity in organs and tissues are given in the works by Ilenko and Fedorov, there is no reason to doubt the high levels of contamination they cite. Consequently, either their territory had nothing to do with the territory contaminated "experimentally" by Korsakov's group, or Korsakov and his colleagues falsified the real contamination data in order to make their work look experimental.

In his summary article (27), Ilenko repeats the data from his first article (10) on radioactivity levels in fish (roach, pike, etc.)—data which we have already analyzed. Those measurements were made in the first year, the summer of 1969. Since the analysis of radioactivity in animals was conducted in 1969 on an expanding scale, it is evident that *Lake X, concerning which various suppositions were made in the previous chapter, was located in the same territorial massif as the sections of woodland, meadow, and other biocenoses contaminated with strontium and cesium. Since these areas ran to dozens of square kilometers, it is most likely that the surface runoff and ground water from this territory were the source of the variations in radioactivity in the lake water.*

At the same time that this extensive experimental work was going on, requiring large inputs of time and a large number of technical personnel, A. I. Ilenko and A. D. Pokarzhevsky were conducting yet another series of studies (29) on how *different biocenoses* influence the concentration of strontium-90 in small mammals (five species of mice and shrews). This work was printed in 1972, but the research was done in June-July 1968 and June-August 1969. (June-August 1969 was the time, as we have seen, when research on strontium and cesium in the fish of Lake X began [10].) In the new publication it is reported *for the first time* that "the contamination of the areas occurred several years before the beginning of our research, and at the time our work was done strontium-90 had been completely assimilated by the biocenoses and had become part of the regular cyclical interchange of substances" (29, p. 1219). At the same time, *this was another new territory,* with quite different strontium levels. Five separate areas, or sectors, were selected in this territory, with contamination levels of 3.21, 1.23, 0.44, 0.37, and 0.14 microcuries per square meter. The fact that the contamination was in microcuries meant that the level of activity in the soil was 1,000–2,000 times lower than in Ilenko's other research. In regard to these contamination levels, see chapter 7 below. To judge from another work carried out in the same sectors, the dosage was incorrectly or mistakenly given in microcuries. In fact these were millicuries. But the research aims were basically the same. The sectors had different kinds of soil and differing vegetation, and the levels of radioactivity varied. Once again this is evidence of *accidental contamination,* for a truly experimental comparison of the different ecosystems would have required equal levels of contamination in each sector. The authors give a detailed botanical description of the areas used, indicating the dominant plant species, the geographical relief of each locality, etc. Although the sizes of the sectors are not given, it is evident from the descriptions that they were rather large, no less than five to ten hectares each, at the minimum. One area (sector 3) was "*on the shore of a lake.*" The variations in soil type (gray forest soil in sector 1, sandy slope in sector 3, and black earth [chernozem] in sectors 2, 4, and 5) testify to the *geographical* magnitude of the total zone. Moreover, the sectors had to be fairly far apart to prevent migrating species (shrews) from crossing from one to another.

If we assume that in the particular region where Ilenko and his associates did their work a rather extensive industrial contamination was present, having occurred several years before the studies on mammals began, we might conclude that these five sectors, with much lower radioactivity than in other works by the same group, were located on the periphery of the main contaminated zone or were even the result of "secondary contamination" (28), caused by wind-borne dust, pollen, or other matter. There is an especially marked occurrence of the spread of radiation by dust, as the work by Korsakov et al. shows, in the spring and late fall, when there is neither foliage nor snow cover.

(It must be noted, nevertheless, that as we shall see in chapter 7, the same sectors appear in later studies of birds, but with contamination levels given in millicuries, not microcuries. In Russian tests this mistake appears quite frequently. In the early 1950s the internationally accepted abbreviations for radioactivity terms were transliterated into Russian with *mCi* [millicurie] becoming МКЮРИ , and *μCi* [microcurie] becoming МККЮРИ. Two Кs in a row for "microcuries" would often make the typist, proofreader, or typesetter think there was an error, and a К would be dropped. Or if МКК was repeated many times, the sudden appearance of an МК might be thought an error and be changed to МКК. In reading the proofs, an author could easily miss this small but very essential misrepresentation.)

Chapter 6

Identification of the Contaminated Zone as the Chelyabinsk Region and the Time of the Disaster as Fall-Winter 1957

In the preceding chapters I have discussed the Urals as the site of all these studies, mainly on the basis of indirect evidence. Above all, the mixture of animal species is typical of the central and southern Urals regions and Western Siberia (see 25—the whole volume—and 15, 30, 31, 32). The plant species likewise represent a mixture of Western Siberian, Urals, and European species. All this evidence points to an area between Europe and Asia in the climatic zone of the southern Urals (several steppe species are mentioned and there are black-earth soil areas as well as those with forest soils and other types of soils). The southern Urals region has low precipitation and this is reflected in both the vegetation and soil described in these studies. But only an approximate geographical area can be established on the basis of the habitats of the various plant and animal species; it cannot be pinpointed more precisely. And of course none of the previously cited works or any that we will deal with—*with one ex-*

ception—give the geographic location of the areas studied, in contrast to accepted standards followed in all foreign publications on radioecology. Yet *ecological principles* strictly require such information. Many biological questions depend on the precise locality, the climate, and similar factors. Strontium-90 and cesium-137 are absorbed by plants and animals in quite different ways in different geographical areas. Precipitation levels, soil type, soil composition, temperature, the length of winters, and many other geographical conditions affect the absorption process. The absence of such data lowers the value of these studies, but apparently the location of the contaminated areas could not be given for reasons of censorship.

However, Ilenko and his associates finally indicated in a relatively recent publication (33) that this part of the Urals was the area in which they caught the animals for their research—although even in Ilenko's book (12) he refrained from giving the location of the studies. Apparently this revelation was the result of an oversight by the authors and censors.

In this paper the research aim was to investigate the possible adaptation of several species of mice (the same species as before) to the radioactive background as a result of prolonged habitation in a biocenosis contaminated with strontium-90. The problem is a rather interesting one and was also considered by geneticists in some studies that I will discuss below. The question posed by Ilenko and his associates was whether *a more radioresistant strain of mice would be selected out* after long habitation in the radioactive environment. (The higher mortality rate of mice in the contaminated areas was demonstrated earlier by the same author, and so the question of the possible selection of such a strain arises logically.)

To answer this question, mice caught in both radioactive and "clean" areas were subjected to additional external radiation at various dosages. It was expected that mice populations which had gone through many generations in the radioactive area would be less sensitive.

In such a work it is obligatory to indicate how many years the population had existed in the radioactive biocenosis.

In this case, mice of various species were caught in several "clean" control areas and in areas contaminated by strontium-90 at levels of 0.2 and 1.2 millicuries per square meter. Once again, these levels differ from previous ones, but the work was carried out several years later and the level of radioactivity could have changed. (Doses of 1.8–3.4 millicuries per square meter were found in the 1964–1965 experiments.) A report on the new work states that "in the fall of 1970 and of 1971 special research was carried out on a group of red field mice and common wood-mice taken from a rodent population which had lived for fourteen years in areas artificially contaminated with strontium-90" (33, p. 573). The authors indicate that *"fourteen years" refers to the beginning of their research, that is, the fall of 1970. Consequently, the contamination occurred in the fall of 1957*. It should not be forgotten that in the fall of 1970 Ilenko was continuing his intensive work on radioactivity in Lake X (11).

The mice were caught "in the outskirts of Moscow and in the Chelyabinsk region." Two of the six species were caught in parts of the central Urals (the Sverdlovsk region). The irradiation of the mice caught in the Chelyabinsk region and in the central Urals was carried out after the mice were delivered alive to Moscow, where there were special facilities for external radiation and a vivarium for observations on the mortality-rate patterns after exposure.

During the fourteen years the mice had lived in the contaminated area, as the authors write, *"there were more than thirty generations of rodents"* in the area. Weak adaptational changes were observed in one species; in another there were none. That is not what is important. It is true that over fourteen years there will be thirty to thirty-five generations of mice. But although the individual adult will not migrate very far from its nest in search of food, the offspring, after growing up and becoming self-sufficient, will presumably migrate greater distances from the birthplace. A study of the distances various species will migrate (24) shows, for example, that the red field mouse, which feeds on seeds and is highly mobile, will go as far as five hundred meters from the nest where it was born. In other species, *new litters* may migrate distances as great as a thousand meters or more. In seasons of low food supply *mice migrations may cover tens of kilometers.* Over a fourteen-year period hungry seasons were bound to occur. One study (25) gives the yearly population dynamics for woodmice in the Urals, showing that there were sharp variations caused by food shortages affecting red field mice. In attempting to determine whether there had been *adaptation,* Ilenko had to be absolutely sure that all of the thirty to thirty-five generations inhabiting the radioactive territory had really lived in that area, even allowing for the scattering of new litters and migration in search of food. In this case the minimal radius for such a zone over thirty generations must be no less than thirty kilometers, if we accept the stated research aim as the true one. Consequently, the contaminated zone for each of the two levels of radiation must have been *no less than ten to fifteen hundred square kilometers.*

It is certainly not possible to compare radiosensitivity in mice from the radioactive biocenosis in the Urals with "control" mice from the Moscow region, because the sharp differences in climate and other factors, producing different races of mice within the same species, would interfere. The "Moscow mice" apparently were only used to compare differences in radiosensitivity between species. In that case, to judge by the table given by the authors, six species of mice were compared. Only two species from the radioactive environment in the Chelyabinsk region—red field mice and woodmice—were studied comparatively. Common field mice and baby mice were caught in the central Urals (the Sverdlovsk region), and these were also part of the group intended for a comparative study of radiosensitivity, not of the group living in the radioactive biocenosis.

To sum up all the data analyzed thus far, we can give a pre-

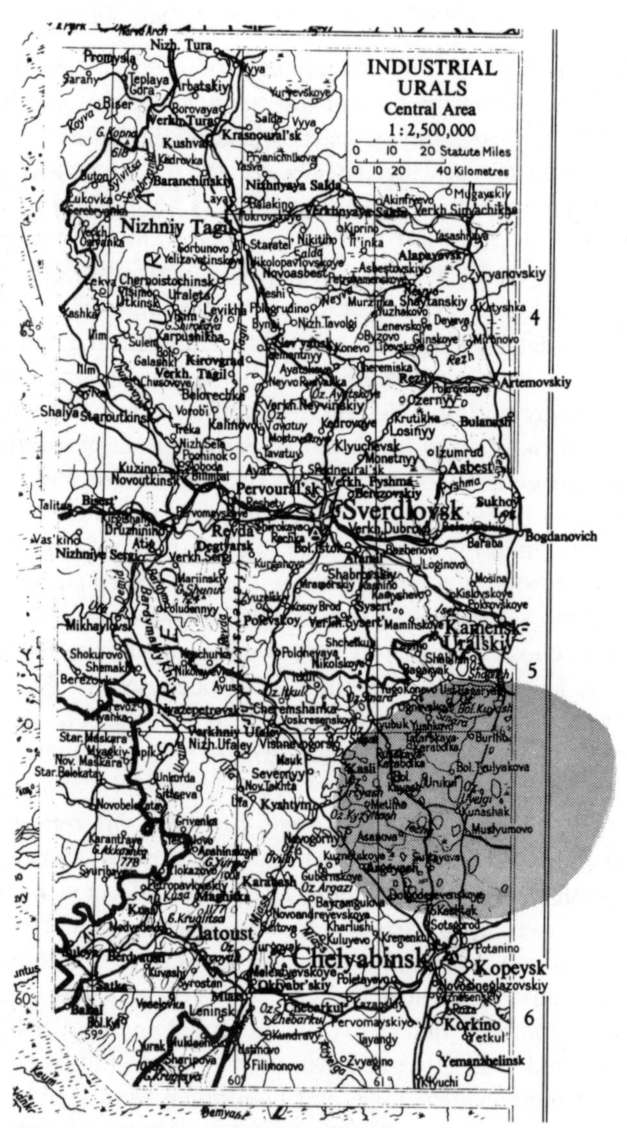

Figure 4. Main zone of contamination.

Source: *The Times Atlas of the World,* vol. II, Southwest Asia and Russia, plate 47. Shaded area (added by author) indicates approximate contaminated zone.

liminary description of the research area discussed in chapters 4, 5, and 6. The area is located in the Chelyabinsk region, covers an area of no less than fifteen hundred square kilometers and includes several lakes. The radioactive contamination of this area with strontium-90, cesium-137, and a smaller quantity of other isotopes occurred in the fall of 1957. The contamination level is measured in millions or tens of millions of curies. From this main region of contamination, with activity varying between one and four millicuries per square meter, a secondary spread of activity through soil erosion and dust occurred in various directions, creating radioactive areas in part of the neighboring Kurgan region and the southern part of the Chelyabinsk region (chernozem soils). But in these areas the contamination level was much lower.

Apparently evacuation of the population also occurred in the areas of "secondary" contamination and these areas became "free" for radioecology studies. But where there were very weak levels of "secondary" or "tertiary" contamination the population was not evacuated, which made it possible for some authors (Korsakov et al. [28]) to make observations on the distribution of radioisotopes in a "human agricultural system."

Chapter 7

Birds in the Radioactive Biocenosis and the Spread of Radioactivity to Other Countries

I have already mentioned two studies (26, 27) in which measurements were made of strontium and cesium levels not only in mammals but in birds inhabiting the radioactive biocenosis. Birds are a special group in the food chain. Some species feed on the seeds of trees and field plants, some on flying insects, some on crawling insects; some on animals, some on small mammals, and some even on carrion. Water birds eat fish, amphibians, and many other "products" of lakes and ponds.

Any comprehensive study of the flora and fauna of the radioactive biocenosis would have to include birds. But in the spring many species of birds migrate to the USSR from farther south, and in the fall they return to winter in such areas as the Mediterranean, Central Asia, Georgia, the Crimea, North Africa, and Iran, where they remain for many months. Not all of them return—20–30 percent are likely to perish from various causes. Moreover, not every bird returns to exactly the same place it left the preceding autumn. Such features of migration as conservatism and flight paths are studied by various methods, primarily the tagging method.

The "hot zone" of dangerous contamination in 1957 could not have been larger than two thousand square kilometers, but the hunting of birds was prohibited throughout the central and southern Urals. Cesium in the muscles of some groups could have reached high levels in the first few years, but strontium in the bones must have been very dangerous too. The bones of young birds are often eaten with the flesh.

However, in other countries and in the southern parts of the USSR where the flocks migrated from the Urals in the winter, hunting was not banned. I do not think that the people in those areas were overly endangered by eating, let us say, a wild duck from Lake X. Nevertheless, the Soviet government should have had an international program of observation and warning in regard to the flights of radioactive birds. It seems that no one thought of this at first, and afterwards they decided it was too late. If radioactive birds did not die during the summer or from their long flight south, it follows that hunters who brought such birds home to feed their families would not die either. Birds have one of the most highly active types of erythropoiesis, or red blood cell formation, and they are among the most radiosensitive of animals. As for the "late" effects of radiation, almost nothing was known of them in the USSR in the years 1958–1963. This was a subject treated by "chromosomal" genetics, but it was banned at the time.

It was Ilenko again who in 1970 published the first detailed study of birds in the radioactive biocenosis (34). There was no separate "methods" section in his article, but in describing the nature of the contamination, he wrote: "Birds were shot in areas artificially contaminated with strontium-90 at levels of 1.8–3.4 millicuries per square meter and with cesium-137 at levels of 4–8 microcuries per square meter, in imitation of an industrial contamination."

As we can see, this is the same biocenosis (with 1.8–3.5 millicuries per square meter) in which mice were caught earlier. However, its radiation levels were dangerously high, and the mice living in the environment had shorter life spans and displayed anatomical and physiological changes resulting from the high radiation level. It is well known that if the contamination level in a territory greatly disturbs the normal biological functioning of animals, that area is hardly suitable for the study of ecological interrelationships.

The table presented by the author shows strontium-90 in microcuries per gram of bone tissue and cesium in mircocuries per kilogram of muscle tissue. There were very great variations in concentration levels between species, but they were expressed in a rather peculiar form, wet weight divided by 100 ($X\ 10^{-2}$). This system of measurement is not very common—it is not clear to me exactly which figures were divided by 100, or why. If it was appropriate to divide the figures in the table by 100, it would have been easier to give them in picocuries, that is, units 1,000 times smaller than microcuries. Picocuries are the usual unit in radioecological studies. But apparently the data obtained by the authors would have been abnormally high if given in picocuries, and would have immediately attracted attention. For professonal radioecologists, I reproduce the table here in full (Table 1). The strontium concentration in the bones of some species is certainly extremely high and could, over time, cause radiation damage to the marrow, the most radiosensitive bone tissue.

This table and the list of Latin names of other species of birds cited in the notes as having been studied in further research may be of interest to ornithologists who would wish to trace the migratory species and investigate whether there are birds of these species with strontium-90 in their skeletons, say, in the Nile Valley or on Greek and Yugoslav islands. Of course not all of the twenty-one species in the table are migratory. Some winter in the Urals; others do not migrate very far—that is, to

Table 1. Concentration of strontium-90 (with yttrium-90) and cesium-137 in the bodies and food of birds in a forest environment (X 10^{-2})

Species	Strontium-90			Cesium-137		
	Number of specimens	Concentration in skeletons	Content in food	Number of averaged samples	Concentration in muscles	Content in food
1. Great tit (*Parus major*)	6	9.0	6.2	—	—	—
2. Willow tit (*P. montanus*)	—	—	—	1	0.41	—
3. Willow warbler (*Phylloscopus trochilus*)	5	25.4	0.22	1	1.44	—
4. Tree pipit (*Anthus trivialus*)	11	2.5	0.63	9	1.8	—
5. Yellowhammer (*Emberiza citrinella*)	6	11.0	0.61	6	0.62	1.5
6. Yellow-breasted bunting (*E. aureola*)	1	5.7	0.54	4	1.2	—
7. Tree sparrow (*Passer montanus*)	7	9.2	0.41	6	2.0	2.3
8. Starling (*Sturnus vulgaris*)	5	2.2	—	3	1.8	9.0
9. Fieldfare (*Turdus pilaris*)	—	—	—	12	0.38	0.85
10. Golden oriole (*Oriolus oriolus*)	5	0.27	0.046	3	0.086	—
11. Scarlet grosbeak (*Carpodacus erythrinus*)	—	—	—	1	0.85	—
12. Garden warbler (*Sylvia borin*)	—	—	—	3	1.74	—
13. Magpie (*Pica pica*)	2	9.0	—	3	6.3	8.1
14. Crow (*Corvus corone*)	—	—	—	3	0.36	3.0
15. Wryneck (*Jynx torquilla*)	4	27.0	0.37	—	—	—
16. White-backed woodpecker (*Dendrocopos leucotos*)	3	8.2	0.19	—	—	—
17. Nightjar, or goatsucker (*Caprimulgus europaeus*)	2	0.08	0.005	—	—	—
18. Black grouse (*Lyrurus tetrex*)	4	5.4	0.12	11	0.55	1.1
19. Great kestrel (*Falco tinnunculus*)	—	—	—	2	0.054	0.1 †
20. Tawny owl (*Strix aluco*)	2	2.1	19.0*	4	1.2	2.1 †
21. Long-eared owl (*Asio otus*)	—	—	—	3	0.9	2.0 †

* Concentration of strontium-90 (with yttrium-90) in the skeletons of small rodents and birds (127 specimens) caught in the area where owls were shot.

† Concentration of cesium-137 in the bodies of small rodents caught in the hunting territory of birds of prey.

From Ilenko (34).

the Caucasus, the Caspian or Black Sea, Turkmenistan, or Uzbekistan. The paper does not give the dates of the dosimetric measurements, but the data themselves make it clear enough, for the first time, that the territory was *contaminated simultaneously by strontium and cesium.* The strontium levels are exactly the same as in two other works (20, 21), and the cesium levels are identical with those indicated in a previously cited paper by Ilenko and Fedorov (26). Once again, there is a reference to Korsakov et al. (28), but as I have pointed out, this is undoubtedly a falsification. In Korsakov's work the contamination was a thousand times lower and the contaminated area included *towns and villages,* for which the doses in the works by Ilenko would have been too dangerous.

One circumstance is very peculiar. In Ilenko and Fedorov's paper (26) seven species of birds were taken for cesium analysis (magpie, starling, black grouse, tree sparrow, tree pipit, yellowhammer, and tawny owl). This must have been the same research operation, because all the figures involving birds from Ilenko and Fedorov's paper are repeated in the article by Ilenko (34) published the same year. However, the article gives data for twenty-one species, not seven, and both strontium and cesium were measured, not just cesium. For four of these species in the part having to do with cesium (the concentration level in the skeleton and in food) the figures are absolutely identical. The paper by Ilenko and Fedorov (26) tells precisely when the birds were shot—June 1968—and thus we must assume that the cesium concentrations of 4–8 microcuries per square meter relate to 1967–1968. But the strontium concentrations of 1.8–3.4 microcuries per square meter (20, 21) were measured in 1964! By 1968 these figures must *necessarily* have been different, because four years could not pass without environmental changes in the strontium concentration. Thus it is possible that no actual dosimetric monitoring of the environment was done after 1964 and the author had to take figures that were old and inappropriate. But it may also be that the researchers moved the "experimental" zone closer to the center of the contamination, where the levels might have been the same even four years later.

In this initial phase of the research about two hundred birds were shot. To do this in June without disrupting the ecological balance, several square kilometers would have been required. According to two bird atlases (35, 36), all of these species are found in the Urals, most of them are typical of mixed forest areas, but some are more typical of agricultural areas, even of towns. Ilenko states that all the species were shot in a forest biocenosis. But four species (*Emberiza citrinella, Emberiza aureola, Sturnus vulgaris, Corvus corone*) are not typical forest species; they prefer sparse mixed forest and open areas. The sparrow, *Passer montanus,* usually inhabits populated agricultural areas, not just fields in general. Once again, this testifies to the great extent of the territory.

Of all these species not one is typical of watery areas or open bodies of water, which might seem unusual considering the fact that there are so many lakes in the southern Urals area. This puzzling factor, however, was removed four years later when Ilenko and I. A. Riabtsev published a special study on *water birds* (37), the purpose of which was very original and dealt with the *migratory conservatism* of birds that return in the spring to the same lake left in the autumn.

The research method was very simple. Birds inhabiting the radioactive lake had strontium-90 in their bones. If there was strontium in the bones of birds caught or shot in the spring, it was assumed they had returned to their previous habitat. "Clean birds" were assumed to be "immigrants," new arrivals from other places. Strontium is especially appropriate for such experiments, because cesium, concentrated in the muscles, would be lost in the southern areas where the birds wintered.

The best way to indicate the methods and results of this interesting work is to quote it directly:

> The present work was carried out on an experimental pond artificially contaminated by strontium-90 (Rovinsky 1965).

This isotope becomes firmly fixed in the bone tissue and is not eliminated for quite a long time. The degree of nesting conservatism was determined from the number of individual birds having strontium-90 in their skeletons and from the amount of this radioisotope in the bone tissue of these birds. These tracers made it possible to identify birds which had previously nested or matured on the body of water being studied and reflect the extent of contact with the original nesting territory in the early spring of 1970–72, at the time of their return from winter habitats. The shooting of the birds took place at temporarily uncontaminated bodies of water before the thawing of the contamined pond and on that pond itself after it was free of ice. The birds shot were primarily drakes.

In the areas of our research, for many years there had been no anthropogenic effects on animals (e.g., hunting, fishing, the burning of meadows, the mowing of hay, and the pasturing of livestock), which is very important for the formation of local populations of water birds.

The following table appeared at this point in the text:

Number of "contaminated" ducks of various species before and at the beginning of the nesting period.

Species	Males		Females	
	Number taken	Those with Sr^{90}	Number taken	Those Sr^{90}
Mallard (*Anas platyrhynchos*)	18	11	7	7
Gadwell (*Anas strepera*)	26	19	13	13
Teal (*Anas crecca*)	11	5	4	3
Garganey (*Anas querquedula*)	9	1	3	3
Pochard (*Aythya ferina*)	6	4	6	6
Total	70	40	33	32

The text continued:

> The number of drakes and ducks with strontium in their skeletons is not the same (see the table). The overwhelming majority of females had the radioisotope in their skeletons, whereas drakes with the isotope in their bone tissue varied from 11 to 73 percent of the total number studied. Drakes without strontium-90 in their skeletons had undoubtedly come to this body of water for the first time (37, p. 308).

The work with birds began *in the spring of 1970*. That same spring Ilenko was still sampling the water of Lake X for radioactivity every month, catching pike and roach there, and taking samples of the biomass and algae. At the same time, work on other projects was also under way. There can be no doubt that all of this simultaneous research was done in adjoining areas and that a large number of technical personnel were engaged in it. (It is not customary in the USSR to include laboratory personnel and other technicians as coauthors in research papers, but sometimes the more courteous scientists will list acknowledgments "for technical assistance." This was not the practice of Ilenko and his associates; they would have had too long a list of names.)

We may ask why the birds in this study were taken from the lakes studied earlier by Rovinsky (9) and not from Lake X? The fact that Lake X is different from Rovinsky's lakes is evident from the behavior of the radioactivity in the lakes. There were sharp variations in Lake X (a running-water lake); but in the lakes we learned about from Rovinsky's 1965 study the activity had been stable since 1960 (see Figure 1). This very fact is undoubtedly what made it possible to "recognize" returning birds by the more or less precise strontium level in their skeletons, whereas on Lake X there would have been great variations in the amount of radioactive isotope in the birds' skeletons.

Rovinsky, as I pointed out, did not give absolute figures for the contamination levels, but cited relative magnitudes in the form of logarithms. He made observations at two lakes, each of geographically significant size (11.3 square kilometers and 4.5 square kilometers). Ilenko calls one of these lakes a "pond"—but this is evidently for the censors. What kind of "pond" can you shoot more than a hundred waterfowl on? However, Ilenko and Riabtsev, in contrast to their own principles as reflected in their previous works, do not give any figures (even in microcuries) for the strontium content in the bones of the birds. In other papers on birds (20, 21, 26, 34, 38), the absolute strontium levels in the bones were always given. Ilenko and Riabtsev *were making readings for strontium-90 in birds from this body of water, but the table only registers the presence of strontium-90, not its level.* This reduces the value of their work as far as strictly *"local"* conservatism is concerned, because Rovinsky studied *two* lakes, not one. Moreover, there was a third lake in the same geographical area with strontium-90 present; this is known from the published data. How could the authors *recognize the birds returning precisely to their "own" lake, the one from which they had flown in the previous autumn? With the proper scientific approach this could only be done by the level of strontium in the bones, which could vary in birds inhabiting different lakes. The radioactivity level would have to be measured before they flew away and after they returned. And there was time to do that, since the research went on from* 1970 to 1972. My hypothetical explanation is that the level of strontium-90 in the bones of the birds on these particular lakes *was very high.* This prevented Rovinsky from stating the actual figures in 1964, when he submitted his work for publication, and it prevented Ilenko and Riabtsev from giving absolute figures. Even ten years after the "plateau" of the curve in Figure 1 had been reached the contamination level of the lakes was too high.

There is another puzzling aspect to this study, namely that the birds were shot "at temporarily uncontaminated bodies of water before the thawing of the contaminated pond and on that pond itself after it was free of ice." *Why were the birds shot on bodies of water farther to the south?* (The fact that they were farther south is evident from the fact that they were free of ice earlier than in the Chelyabinsk region.)

In the Chelyabinsk region there are hundreds of large and small lakes; the total area covered by such bodies of water in this region is more than sixty thousand hectares. The contamined zone is located north of Chelyabinsk, but there are a great many large and small lakes south of it as well. There is no guarantee that birds from the lakes south of Chelyabinsk would fly farther north. How were the birds which landed there only temporarily (on their way back to "Rovinsky's" lakes, for example) distinguished from those returning to their "own" lake south of Chelyabinsk to remain for the summer? The ratio of "dirty" to "clean" birds on the lakes south of the dirty zone must have differed from the ratio in the dirty zone itself. There are also a great many lakes farther north, in the Sverdlovsk region and other areas, and the lakes in the dirty zone must have been temporary stopping-places for birds still on their way north.

I can only advance one hypothesis to explain these methodological anomalies. Apparently there is a program that is not at all scientific, aimed at the systematic destruction of the migratory birds which inhabit the heavily contaminated areas. The shooting is done in the spring, summer, and autumn—in order to reduce the dispersion of radioactivity to other parts of the USSR and other countries. It is impossible to totally eradicate migratory birds, because they do not have absolute nesting conservatism and each year the lakes in the "dirty" zone are repopulated. But if this is the case, it is meaningless to present the problem as a scientific one.

The authors unintentionally confirm the reports that hunting and fishing had been banned in the area of their investigations *for many years,* that is, that this was an evacuated zone, where there had previously been both livestock and the mowing of hay and consequently a settled human population as well.

The authors avoided publishing any real indications (in microcuries or counts per minute) of the radioactivity level in the bones of various species. They did, however, publish data from a comparison between the levels of radioactivity in males and in females. These results are given in "relative magnitudes," bars in a graph. In all the species studied, the strontium-90 concentration was greater in the females, with the disparity being very great (ten to fifteen times) in three of the species. Ilenko and Riabtsev express the opinion that "this phenomenon evidently has to do with the fact that the females remain longer in the nesting territory contaminated by the isotopes. The drakes leave the breeding grounds first. Late in the spring they migrate to intermediate areas."

This explanation does not seem convincing to me. It might account for small differences, but not a quantity ten to fifteen times larger than another. The lakes studied by Rovinsky were contaminated in 1957 or 1958. As an isotope deposited in the bones for many years, strontium-90 accumulates for several years until it reaches a certain "plateau," and this "plateau" may have been reached by birds that included this lake in their "itineraries" for many years. There could only be large differences in the case of very young birds, which would evidently be typical of this area if there was systematic shooting of animals contaminated by strontium. Such systematic shooting of migratory birds could be justified since it prevents the serious spread of dangerous radioactivity. However, this makes properly planned scientific experimentation impossible. We can see once again why the conclusions in the papers we are citing would be more interesting if the authors had not been obliged to conceal so many of the real conditions in which their research was done.

In 1975 Ilenko, Riabtsev, and D. E. Fedorov (38) published a more extensive paper on the same problem, but for migratory "land birds." The research aim—to test for conservatism in birds returning from winter habitats—was applied to sixteen species inhabiting an area contaminated by strontium-90. The birds were shot in the same five sectors described earlier in the work by Ilenko and Pokarzhevsky (29), published in 1972. Either in the 1972 work or in the one on birds published in 1975, a serious error was made in specifying the contamination level in the territory. The 1972 paper gave the figures for the contamination per square meter (3.21, 1.23, 0.44, 0.37, 0.14) as *microcuries.* The paper published in 1975 refers to the same figures as millicuries, that is, a contamination level 1,000 times greater. Through a comparison between the data for strontium-90 in the food and skeletons of two similar species (tree pipit and yellow-breasted bunting), one in the 1975 paper and one in Ilenko's first paper on birds (34), where the contamination is given in millicuries (1.8–3.4), the conclusion may be reached that the mistake was made in the 1972 publication and that the contamination in the five sectors should be given in millicuries. This does not change our conclusion at the end of the section on mammals as to the existence of zones of "secondary" contamination, because the contamination levels in sectors 4 and 5 were still much lower than in most of the studies of the main contaminated area.

The unacceptable dosimetric methods used in all this research is evident from the simplest fact. In 1972 all of these figures (3.12, 1.23, etc), as well as such precise calculations as $0.14^{\pm}$ and 0.001, were said to apply to the habitats of the animals caught and shot between *1968 and 1969*. In the case of birds, it is clearly indicated that they were shot from *April to June 1973*, that is, five years later! But the environmental radioactivity should have declined by approximately 10 percent during this time through the decay of strontium alone. And in addition to radioactive decay, there are processes of absorption, various types of erosion, etc. How can five-year-old figures be given without any correction even for radioactive decay? This elementary methodological illiteracy only emphasizes how careless the dosimetry was in all this research.

A total of 469 birds of sixteen species were shot for these studies. The list of species with their Latin names is given below* for those specialists who might wish to seek out radioactive birds in their winter habitats in various regions outside the USSR. Each species has its own particular flight path. Some travel great distances to Iran, Turkey, and North Africa. Others winter on the northern shores of the Caspian or Black Sea or in Soviet Central Asia. As one might expect, different species of birds have different degrees of nesting conservatism (the tendency to return to the same places after winter migration). For conservative species, the percentage of "immigrants" is not very large (14–25 percent). For the less conservative, the "immigrants" from other areas may be as much as half the population.

The idea of a study of the accumulation of strontium-90 and cesium-137 in the organisms of various species of birds at different times of year and in relation to different types of food had already been carried out much earlier, in 1959, in the contaminated biocenosis near the Oak Ridge National Laboratory in Tennessee (39). In that case the soil was contaminated much more heavily by cesium-137 than by strontium-90. Tennessee has a milder climate than the Chelyabinsk region, and the population "density" of birds per hectare (or in the United States, per acre) is evidently higher in the central United States. According to the data in this paper, the density of the bird population was twenty per acre. Ilenko and his associates shot about five hundred birds for only one of their experiments, with by no means all of the species typical of the area being shot and only part of the population of each species. In the five sectors chosen by the researchers there were thousands of birds, and each sec-

* Migratory birds studied by Ilenko et al. (38) in the radioactive environment: tree pipit (*Anthus trivialis*), yellow wagtail (*Motacilla flava*), yellow-breasted bunting (*Emberiza aureola*), yellowhammer (*E. citrinella*), reed bunting (*E. schoeniclus*), chaffinch (*Fringilla coelebs*), scarlet grosbeak (*Carpodacus erythrinus*), golden oriole (*Oriolus oriolus*), stonechat (*Saxicola torquata*), whinehat (*S. rubetra*), fieldfare (*Turdus pilaris*), Siberian brown shrike (*Lanius cristatus*), great tit (*Parus major*), willow tit (*P. montanus*), and rook (*Corvus frugilegus*).

tor was separated from the others by a considerable distance. All this testifies once again to the existence of tens of square kilometers of contaminated territory.

Of course there are dozens of other species of migratory birds in the central and southern Urals region, and their fate can be studied outside the Soviet Union. Conservatism in the winter migration patterns could also be studied. But this cannot be done without cooperation by Soviet ornithologists and without assurance that an annual program of "prophylactic" shooting of birds in the contaminated areas is not being conducted.

Chapter 8

Soil Animals in the Urals Contaminated Zone

Among the hundreds of different papers on plant and animal radioecology published in Soviet journals and books since 1965, it is relatively easy to pick out those having to do with the Urals radioactive zone. In research where a radioactive microecology was artificially created, all the necessary methodological details are described with care, the contamination levels are comparatively low, and there is a certain regular pattern to the contamination. The radioisotope mixture used in such experimentation is more heterogeneous and often includes short-lived fission products. In many cases, the geographical location of the research is indicated.

Experiments done in the Urals zone are usually "tied" to the same radioactivity levels we have encountered in works by Ilenko and others. The assertion can almost always be found that the contamination was introduced "for experimental purposes" anywhere from six to fourteen or more years before the research began. Subtracting this number of years from the date of the experiments we usually arrive at late 1957 or early 1958. If the "experimental contamination" seems to date from a later time, the activity level is usually much lower and has the characteristics of "secondary" contamination. The location of the research is not specified. And the experimental fields are not described, reference being made instead to earlier publications by Ilenko or other works we have cited in chapters 4–7.

It was to be expected that as soon as the contaminated region in the Urals was opened up for ecological research, scientific institutions of all kinds would take advantage of the opportunity. Studies lasting many years would begin not only on fish, mammals, and birds, but on all plant and animal groups. And that is what happened. I do not propose to review in detail *all* the scientific findings obtained from this unique ecological zone. A large academic monograph might be devoted to that subject. Mine is a more limited purpose—*to point out exactly what was not made public* in the widely ramified radioecology studies on fish, mice, rabbits, reindeer, birds, mosquitoes, ants, frogs, plants, etc. This aim has determined my selection of material on which to comment.

Sometimes a publication that outwardly seems quite ordinary will attract the eye of an experienced observer because of certain peculiar features. As an example, we can take a symposium on radionuclide migration in soil and plants, held in early 1971 in Tbilisi. The report on this symposium published in the *Bulletin of the USSR Academy of Sciences* (40) contains brief abstracts of the papers presented, but it is hard to draw conclusions relating to the Urals zone from any of these. Much of the research reported to this symposium was definitely experimental and contained references to such areas as Byelorussia, the Baltic, Georgia, and others—areas quite far from the Urals. Nevertheless, it struck me as odd that five reports concerning the dispersion and fixing of strontium-90 and cesium-137 in the soil *did not specify the location of the research* and were presented by scientists connected with the Biophysics Institute of the USSR Ministry of Health. A Western reader might disregard this detail, but to a Soviet expert who knows that this is a secret institute in Moscow, under a special administrative board of the Ministry of Health, it speaks volumes. This special administrative board is headed by Lieutenant-General A. I. Burnazian, of the medical service, and operates a nationwide "closed" system of clinics, hospitals, and scientific institutions serving the atomic industry. As we indicated earlier, Burnazian was in charge of the radiobiological prison institute where Timofeev-Resovsky worked until 1955. In 1960 or somewhat later, Burnazian received the Lenin Prize (not reported in the press) for his part in working out methods for the care and treatment of radiation sickness. Why did this institute, under this particular administrative board, suddenly take such great interest in radioisotopes in the soil? And why didn't this symposium include even one report on soil radioecology from the most prominent Soviet center for soil research, the Soil Sciences Institute of the USSR Academy of Sciences?

It was also in 1971 that the first detailed paper (41) on the radioecology of soil insects and other soil animals in the Urals contaminated zone was published. The authors, M. S. Giliarov and D. A. Krivolutsky, had worked in close association with Ilenko, E. A. Fedorov, and G. N. Romanov, to whom they express their thanks for assistance in organizing the research. Their paper has no section on methods. However, the authors state that the research was done in a contaminated field used earlier by Ilenko and Romanov to study the effect of strontium-90 on mice (20, 21). Ilenko and Romanov, as we have noted, used this field from 1963 to 1964, when the contamination level varied from 1.8 to 3.4 millicuries per square meter. Giliarov and Krivolutsky made their comparative study of various soil animals in radioactive and control fields in 1968 and 1969, but they gave absolutely the same data for the activity of the soil. Once again, this indicates that *the biologists were not allowed to make independent dosimetric measurements of the radioactive contamination.* Researchers arriving from "open" institutes were apparently given an old dosimetric chart of the contaminated territory. Everyone must have realized that such a chart required *annual* revision. But the task of territorial dosimetry depended on other, more secret agencies, which did not consider it necessary to make annual corrections for radioactive decay, erosion, and uptake by plants. As outsiders, the researchers felt that it was better to have an old dosimetric chart than none at all. They had no experience of their own in dosimetry, and their research was confined to measuring radiation patterns in various soil animals.

The published data, which is too unwieldy to give here in detail, show that the given contamination levels (1.8–3.4 millicuries per square meter) were highly destructive for soil animals. Predatory beetles suffered least; their numbers in the contaminated area were reduced to only 66 percent of the figure in the control area. Non-predatory beetles, beetle larvae, and other insects that feed on plants (phytophaga) suffered the most; their numbers fell to 56 percent of those in the control area. Soil animals that feed on organic products in the soil (where the highest level of strontium concentration was found)—the saprophages—died out almost completely; their numbers fell to 1 percent of the control group. Taxonomically, the groups studied were Aranea, Mollusca, Lithoblidae, Geophilidae, Lumbricidae, and Diplopoda.

In all cases, the greatest amount of destruction was observed in animals inhabiting the upper layer of soil, obviously the most severely contaminated one. Data on the strontium concentration at different soil levels were to be expected in such a paper, but the zoologists were not allowed to present such information.

The almost total destruction of certain groups of soil animals in this zone is evidence that the contamination was too great and *ought not to have been used for experiments studying food chains in various groups of animals.* The entire balance of food relations was unquestionably disrupted by the excessively high doses of radioactivity. But almost all the research on food chains in mammals and birds that we have analyzed above were done in the same ecosystem. This means the researchers had no choice.

The results of continued research on this subject, with some additional observations, were published in 1972 by Krivolutsky et al. in an East German scientific journal (42). Although observations done in 1970 were also included in this article, the stated level of soil activity remained unchanged—1.8–3.4 millicuries per square meter. (The authors sometimes gave microcuries by mistake.)

The effect of radioactive contamination on various species of ticks and mites is studied with especially great detail in the first paper (41). Calculations were made for more than twenty different genuses and species of ticks and mites. Many tick and mite species are, of course, parasitic on mammals and carry certain diseases. Although there was no pattern of severe suppression of tick and mite populations in the radioactive environment, it was natural, nevertheless, to try to determine whether the given level of radioactive contamination reduced the extent to which they infected mammals. Ilenko made such a study during the same summer months of 1968–1969 (43). It was done in the same geographical region, but in areas with rather different strontium levels (from 0.6 to 2.5 millicuries per square meter). Giliarov and Krivolutsky (41) explicitly mentioned their collaboration with Ilenko, and the whole group came to the contaminated region from the same institute in Moscow. One wonders why the population dynamics of ticks and mites in the soil were studied at one level of radioactivity, and the degree of infection of mammals by ticks and mites at another? There is no scientific logic in this.

But for Ilenko, this research was a minor addition to the work he was then doing on mammals in various ecosystems. Making a count of ticks and mites on the bodies of animals was not his main task but merely a practical sideline that would supplement the data of Giliarov and Krivolutsky. The fact that their experiments were carried out under different conditions was disregarded.

The soil zoologists from the Institute of Evolutionary Morphology and Ecology of Animals of the (USSR Academy of Sciences) continued their research in the same region. In a 1974 paper (44), which has a reference to the earlier research (41, 42), the role of soil animals in the migration of calcium and strontium in the biocenosis is analyzed. Such a study required determinations of the strontium at different biological levels, and without a properly organized dosimetric technique it was impossible to carry out that task. Therefore the author, Krivolutsky, collaborated this time, not with the zoologist Giliarov, but with A. D. Pokarzhevsky, who (like E. A. Fedorov) has only an indirect affiliation with the Institute of Morphology in Moscow. Once again, there is no methods section in this detailed paper, only a reference to the earlier papers (41, 42), which have no descriptions of methods either, and a reference to Ilenko (20, 21). We have already noted that because of the almost total destruction of some genuses and species of soil animals, this region was not suitable for properly planned ecological research on the *migration* of strontium. All feeding patterns were disrupted.

The migration of strontium and calcium was traced from the soil and soil animals to flying insects, rodents, frogs, large mammals (ungulates, without the species being indicated), and birds (again without indication of species). For each group of animals, summary figures were given for the accumulation of strontium-90 and calcium, with the biomass and the isotope quantity being calculated per hectare. Since ungulates (such as roe deer and reindeer) were included in the experiments, it is evident that the total size of the contaminated zone was rather large. This time, however, despite a reference to the "description of the areas given previously" (41, 42) the authors measured the total radioactivity in various components of the biocenosis. In the soil there were 32 curies per hectare (3.2 millicuries per square meter), but this applied only to the *top 5 centimeters of the surface layer.* Plants contained 5 curies per hectare, fallen leaves and other plant litter, 4.3 curies. For various animals, measurements were given in microcuries, because in this paper the total dry weight of the various species is measured per hectare in kilograms and grams (250,000 kilograms in soil to a depth of 5 centimeters per hectare; 100,000 kilograms in plant material per hectare).

In a footnote to the main table, the authors indicate that the natural decay of strontium-90 in this biocenosis was 0.8 curies per hectare annually. But this may have been just a theoretical figure, a rough estimate. What was the annual reduction in activity from all factors operative in the biocenosis? That was not indicated, because measurements were evidently not made throughout the zone. I deliberately speak of a "zone" with geographical dimensions in mind, because the research work undoubtedly extended far beyond the relatively limited areas selected earlier for the analysis of soil animals alone. The figures of 1.8–3.4 millicuries per square meter (in 41, 42) do not coincide with those given in 1974, and the description of the research area is different. The 1971–1972 papers refer to a "*birch forest* area." The 1974 paper speaks of "forest-and-steppe area"—that is, an ecological zone on quite a different scale.

Some judgments may be made about the size of the zone from the figures in the table of the total strontium content per hectare for various groups of animals. The biomass of ungulates (roe deer) and birds (given only in totals) is calculated per hectare in another paper (45). In dry weight, the "biomass" of ungulates was only 0.3 kilograms per hectare, of birds, only 0.05 kilograms per hectare.

If the dry weight of one reindeer, including horn, bone, and hide, is 50–60 kilograms *at the minimum,* 4,000 hectares would be needed if twenty were shot, as reported by Ilenko (27). But that would mean their complete eradication from the research area, and as we have indicated, Ilenko was studying food chains and presumably would not so drastically alter the population balance. In that case, there should be no less than 90 percent of the original number of animals left in the territory as a whole. That would require 40,000 hectares, or 400 square kilometers. Approximately the same amount of territory would be needed for an "experimental sample" in which 1,000 birds were shot, since birds in one hectare amount to only 50 grams dry weight.

If we calculate the total amount of strontium-90 in the biocenosis being studied, using the data in the main table, we obtain a figure of approximately 43 curies per hectare (4.3 millicuries per square meter). This is much more than the 1.8–3.4 millicuries per square meter cited by Ilenko for his experiments in 1963–1964. Moreover, the soil layer, 5 centimeters deep, accounts for most of the 4.3 millicuries per square meter. It is evident from Figure 1, which shows the distribution of strontium in this area, that the pattern is extremely chaotic. Thus the figures given by Krivolutsky and Pokarzhevsky (44) must have been averaged out. Even so, this average indicates that over the hundreds of square kilometers of this large territory there were, and must still be, particular zones with considerably higher levels of radioactive contamination than stated in the papers we have analyzed.

In another paper (46), Krivolutsky takes a rather insular approach to his data on the effect of radioactive soil contamination on ants, although he states: "This study of the ant population was an integral part of the research in these same experimental fields." The ants were radio-resistant, and their numbers remained virtually unchanged in this birch-forest area with strontium levels of 1.8–3.4 millicuries per hectare. (Five species of ants were studied.) Although all the other work with soil animals and insects in this series examined only the strontium content, the work on ants *measured the effect of cesium-137 and its accumulation in ants.* The area containing cesium-137 is identified as the same one in which Ilenko and Fedorov's work (26) was done. The cesium level was approximately a thousand times lower than the strontium (4.5 millicuries per square meter), but the whole description is worded in such a way as to give the impression that cesium was the *only* isotope in the two-hundred-square-meter contaminated field. We have already seen from an analysis of the materials on mammals that cesium-137 was always associated with strontium-90 in this area, but in much smaller quantities (apparently because it did not fix firmly in the soil).

The paper on ants, published in 1972, was sent to press in 1971. Exactly when the research was done is not indicated. We can assume, however, that the work began in 1968 as part of the general study of invertebrates in the same area. Previous papers failed to indicate either the *total time* soil animals had been exposed to radiation or the date of the original contamination, although these are important from the methodological standpoint. The paper on ants fills in the gap. The authors report that the ants lived "under conditions of *prolonged radioactive contamination* (*about ten years*)." As in the case of mammals, this brings us to 1957 or 1958.

Separate comment is called for on the author's negligence toward, or falsification of, the data on cesium in this paper. The cesium level in this biocenosis is given as 0.005 millicuries per square meter (5 microcuries per square meter). But in a table with data on cesium accumulation in various species of ants, with radioactivity expressed in microcuries per gram of live weight, the figures vary from 3 to 17. Likewise, earthworms had approximately 4.4 microcuries of cesium per gram of live weight, and the earthworm mass per square meter was greater than one gram. There is approximately one ants' nest per meter of forest, with more than twenty thousand ants in each (if the figures cited here are to be accepted). The cesium levels also varied within the nests. In the soil of nest No. 1, there were 11 microcuries per gram, but in the mound over the nest the figure was 50 microcuries per gram. In another nest, the soil contained 150 microcuries per gram, and the mound had 110. These figures are completely out of line with the assertion that the overall contamination was 5 microcuries per square meter. Perhaps cesium was fixed permanently in the anthills in the first years (1958) but was washed out of the soil in the area generally. It is hard to find an explanation. If we add the fact that in the English abstract of the article millicuries and microcuries are mixed up, all we can do is express the wish that the authors of radioecology papers would take a more serious and attentive attitude toward measurements of radioactivity.

We have assumed that all this work was done in the southern Urals on the basis of its close association with the work of Ilenko and others. However, this conclusion may also be drawn from the species dealt with in certain cases, especially the tick and mite species. In the European part of the USSR, for example, the most widely known and frequently studied mite is *Ixodes ricinus,* which infests field mice. This species was not found in the region where Ilenko, Krivolutsky, et al. made their observations. But in the Urals the habitat of this species is limited to the mountains and does not include the southern Urals (47). In the Sverdlovsk region (the central Urals), some isolated specimens are found, but in the southern Urals, not at all. This rules out the European part of the USSR as the site of the experiments. But in the eastern part of the USSR it is precisely the southern Urals that typically has forest-and-steppe zones with birch groves. Farther east, in Siberia, and farther south, in Central Asia, the species of soil invertebrates and plants also change. Through the process of elimination, the three regions of Bashkiria, Chelyabinsk, and Kurgan can be pinpointed as the areas where the zoo-geographic possibilities for such a biocenosis coincide. All three regions are in the southern Urals.

Chapter 9

Trees in the Urals Contaminated Zone

In the preceding chapters we saw that in many cases the catching and shooting of radioactive animals was done in forest zones. Such descriptions as "a birch forest area," "a mixed forest-and-steppe zone," "a concentration of isotopes in the leaf litter and other forest detritus," indicate that there were many forested areas in the contaminated zone. Since trees live for many years, they are more sensitive to prolonged and *chronic* radiation than field plants. This is well known from many studies in radioecology. Many observations were made along these lines at a special gamma field in a forested area at the Brookhaven National Laboratory in the United States from 1960 to 1964 (48). Of the trees, the deciduous genera were much more stable (ten to twenty times) than the conifers, because the periodic loss of leaves prevented accumulation of too high a dose from external radiation or, internally, from the soil. With doses of approximately 2 roentgens per day, pine growth is drastically suppressed, and exposure for five to six years to 6 or 7 roentgens per day kills a pine forest. The so-called absorbed dose (the rad) is the equivalent of a roentgen. If an environment is contaminated, for example, by strontium-90 at a level of 0.1 millicuries per square meter, the absorbed doses over several years amount to thousands of rads (kilorads), and a great many chromosomal aberrations (49) appear in the dividing cells of the growing parts of pine trees. Contamination doses of 2 millicuries per square meter suppress the growth of pine in the very first years and may cause the death of entire stands of trees if they are affected chronically for many years (50). Thus, it is quite obvious that the coniferous forests in the "experimental" areas used for the study of animals over a period of ten to fourteen years, as reported by Ilenko, Krivolutsky, and the others, *must have died out* well before the extensive ecological experiments began in the area in 1964–65. Doses of 1.8–3.4 millicuries per square meter or others within the range of 1–4 millicuries per square meter would be lethal for pine forests, and we could hardly expect this to be mentioned in these publications. As we shall see in subsequent chapters, a group of geneticists headed by Academician N. P. Dubinin worked in the contaminated Urals zone. In his autobiography (51), Dubinin mentions that in 1970 he reported to the Presidium of the Soviet Academy of Sciences on the results of *eleven years of experiments* in forest and meadow areas "which had been contaminated by high doses of radioactive substances. . . . In these circumstances some species died out, some continued to suffer for a long time, their populations becoming reduced in size, and some evolved toward a higher resistance" (51, p. 330).

These facts and admissions show, as I have stressed repeatedly in the previous chapters, that the main doses of contamination in this environment were *too high for the maintenance of a normal biocenosis.* The dosage had a chronic effect on the flora and fauna and caused so much major radiobiological and genetic damage that both the biocenosis and the relations between species were inevitably altered. That is why it is hardly likely that these high doses of contamination were done experimentally, especially since strontium contamination lasts for hundreds of years.

There are several publications on strontium and cesium accumulation in trees which, going by a number of indications, we can assume were done in the contaminated areas of the southern Urals. However, they were either done on the periphery of the contaminated areas or in zones of "secondary contamination," resulting from dust or snow carried by the wind, because contamination levels that would affect stands of trees without seriously harming them over a period of many years' observation would presumably be much lower than the levels we found in the publications on the concentration and distribution of radioisotopes in animals.

Mixed forests are typical of the Chelyabinsk region and if the pine, fir, and other conifer species in the heavily contaminated zones died out between 1959 and 1963 from chronic internal and external radiation, the dead forest areas were undoubtedly cut down or leveled by bulldozers, leaving an area cleared of trees so that bushes and ruderal vegetation, more resistant to radiation, could develop there.

Ecological research on the propagation of isotopes through the components of various forest biocenoses was done with significantly lower levels of radioactive contamination, and certain papers can only be linked with the southern Urals region by the plant species, the timing of the contamination, which can be traced to 1957–1958, or by references to research done in areas previously described by Ilenko. (Ilenko also had fields with varying levels of activity for observations.) Such guesses must be made because, in spite of the *scientific necessity* for indicating the geographical locality of any fairly extensive radioecological research, this is usually not done in the publications of Soviet researchers, especially if it is an area of real, not experimental, contamination. Indicating the geographic locality is an elementary requirement for a scientific article of this type, because with different climatic zones, different levels of precipitation, winters of different lengths, and lack of uniformity in other factors, the radiosensitivity of plants also varies, as does the intensity of the isotope concentrations, the rate at which forest detritus decays, the accessibility of the isotope in the soil, the extent to which the contamination is washed from the surface of the leaves, and many other factors.

If we take for example the study by R. M. Aleksakhin et al. "Peculiarities and Quantitative Prediction of the Cumulative Build-up of Strontium-90 in Woody Plants" published in 1970 (52), we would have every reason to suppose that the research was done in an area peripheral to the Urals contamination, an area in which the level of radioactivity was lower than average.

This paper was presented for publication in 1969. The authors traced the accumulation of strontium-90 in a birch stand (with the average age of the trees being thirty years) and a stand of pine (with the average age being fifty years). It was important for the research that the trees not have been damaged by radiation and, therefore, that the contamination doses be quite low. For comparative work it would be methodologically important that the same concentrations of the radioisotope be applied to both the birch stand and the pine stand. But the strontium in the birch stand was 0.28 microcuries per square meter, while in the pine stand it was 0.015 microcuries per square meter, that is, *twenty times lower!* Measurements of the radioactivity in the birch trees were begun two years after the contamination and completed eleven years later. Measurements were begun in the pine stand only two years after the birch-tree measurements began but were completed in five years. In neither case are the years in which measurements were taken indicated. However, we can assume that if the article was sent to the journal in 1969, the experiments could have been completed either in early 1969 or late 1968. Contamination that occurred eleven years earlier would bring us back to late 1957 or early 1958. Apparently the experiments with pine were begun later in a zone of "secondary" contamination. This would explain why the observations were carried out for a shorter time with such low contamination levels.

Certainly the authors did not contaminate the area themselves in either case, because it is precisely the first years that are most important for determining many things (removal from leaves, fixing in the forest detritus, etc.). In any real experimental contamination of forest ecosystems, all of these detailed processes of the *first year* would have been studied carefully.

In what part of the USSR were these observations made, under what sort of climatic conditions, precipitation levels, etc.? None of these geographically and ecologically important details are indicated. (Two of the same authors, Aleksakhin and Naryshkin, in a number of other works on the radioecology of tree stands refer in the usual way to the location of the forest and provide more details. For example, in a recent book (53), on the distribution of strontium-90 in a fir forest growing in soddy-podzol soil, the authors wrote: "The research was done in the Malinsky division of the Krasnopakhorskoe forestry farm in the Moscow region" (p. 33). But when the work is done on "gray forest soils" and "leached-out black earth soil (chernozem)," and when the Urals zone is indicated by signs already familiar to us, the location of the forests is no longer mentioned despite the far more extensive research effort.

When *serious* mistakes in calculations are discovered after the fact it is customary in the scientific world for "corrections" to be published in the back of a later issue of the same journal in the so-called "Errata" section. In the article by Aleksakhin et al. (52), I suspected a mistake in dosimetry. The contamination level of 0.015 microcuries per square meter was much too low. This level could make it quite difficult for radioactivity to be measured in research work in a forest biocenosis.

However, no corrections to this article were published. Only by accident, in 1977, were my suspicions of an error confirmed.

A special volume summarizing recent data in radioecology was being compiled in the United States. Some material from works by Aleksakhin and his associates was proposed for inclusion in the volume. The Soviet authors were asked for permission to use their data. The response was favorable, but the letter of authorization from the Soviet ecologists, a copy of which is in my possession and which is dated September 8, 1977, contained the following statement:

> [We] agree to the publication of our experimental data in your subject book. We should like to ask you, if possible, to do a corresponding reference to the literary source Aleksakhin, R. M., M. A. Naryshkin, M. A. Bocharova. 1970. Peculiarities and quantitative prediction of the cumulative accumulation of strontium-90 in woody plants. Reports of Academy of Science, USSR 193(5):62–64. We call your attention to the regrettable misprint in the text of our paper in Russian. It is necessary to read (p. 1192), that the quantity of ^{90}Sr on the experimental plot under birch stand is equal not to 0.28 μCi/m^2 but 0.28 *m*Ci/m^2. Similarly the quantity of ^{90}Sr on the experimental plot under pine stand is equal to 0.015 mCi/m^2.
>
> Sincerely Yours,
> R. M. ALEKSAKHIN
> M. A. NARYSHKIN

The "regrettable misprint" represented a factor of 1,000!

It is also peculiar that the article does not indicate the Latin names of the birch and pine species—an elementary rule for ecological research. Different species, for example, of birch, have different rates of growth and undoubtedly have different factors controlling their accumulation of radioisotopes. Apparently the species names would have been too revealing: the locality of the contamination might be detected from those names. In the USSR there are, for example, more than fifty different species of birch; and certain species are peculiar to

Europe, the Caucasus, the Urals, Siberia, and elsewhere. Some species typical of the Urals foothills are only found at certain altitudes above sea level. In the lower part of the hills regions there is *Betula verrucosa;* higher up is *B. pubescens;* still higher, *B. torticosa;* and even higher, *B. humilis* (54).

In the same year, 1970, Aleksakhin et al. published a survey of forest radioecology (55), in which they noted on the basis of data in the literature that the migration of isotopes in the forest "occurs more intensively in the initial, relatively short period after the fallout of radioactive particles." Depending on the climate, there is a period of "partial cleansing" of superficial contamination from the canopy lasting from two weeks to several months. This confirms the view that in the preceding work the authors themselves did not create the radioactive biocenosis but arrived at the scene of the contamination two years after the event.

The peripheral areas of the Urals contamination are apparently used for cytogenetic research on chromosomal abberations in the growing points of pine in areas with contamination varying between 0.1 and 0.15 millicuries per square meter. This research is mentioned in the survey by N. P. Dubinin et al. (49) without any reference to the publication of a paper on the subject giving information on the methods employed. The research was done in a young pine forest containing much more strontium-90 than the pine forest in which Aleksakhin's group worked. But even these levels are ten to thirty times lower than those where Ilenko studied the behavior of animals in a birch forest biocenosis. At levels of 0.1 and 0.15 millicuries per square meter the authors observed a great many chromosomal abberations in the meristem cells of pine and, according to their calculations, the cumulative dose over a number of years was measured in kiloroentgens—therefore high enough to suppress the growth of pine (48). Geneticists and cytologists do not need large areas for their research and it can be assumed that this study was truly experimental from start to finish. However, it is quite unusual and surprising to find the authors acknowledging that "the cytogenetic research was carried out only six years after the trees were dusted with strontium-90" (48, p. 186). It is evident that these researchers, too, arrived at the site of the contamination after a long delay; thus all of their data on contamination levels and absorbed doses are quite arbitrary. Although the dynamics of chromosomal aberrations were studied by season (winter, spring, summer, and fall), neither the species of pine nor the geographical location of the forest is indicated. There is only a passing reference to the fact that the "experiments were carried out in a temperate climatic zone." And yet in reviews and descriptions of the experiments of foreign authors who have done radioecological experiments it is the usual Soviet practice to reproduce both the exact geographical location and the names of the tree species. Moreover, Soviet authors themselves give the species names for herbaceous plants in radioecological publications, because in this case it would be simply absurd to speak of "daisies," "couch grass," and so forth, without the more precise species designation. For forest genera with relatively small natural habitats, the specific Latin name as a rule is not given.

It is interesting to note that Ilenko and other authors, who worked with animals in the extensive area having contamination levels of 1.8–3.4 millicuries per square meter, regarded these levels as acceptable for radioecological research and the analysis of food chains involving plants and various animal species. However, for radioecologists such doses in forest stands are regarded as causing sharp changes in all the relations between species in the contaminated territory. In evaluating the level of radiation damage to forest varieties under conditions of radioactive contamination, F. A. Tikhomirov (50) cites Ilenko's work on mice in a birch-forest biocenosis with activity levels of 1.8–3.4 millicuries per square meter as an example of a *disrupted ecology*. Tikhomirov, as we shall see in analyzing the next paper, also did research in the southern Urals zone. According to Tikhomirov's evaluation, the high radioactivity in the environment Ilenko worked in resulted in suppression of the growth of the birch forest and destruction of young trees. The forest thinned out and as a result there was a severalfold increase in the amount of grass, which led to a big increase in the number of mice. The weakened trees were easily infested by insect pests and this attracted insect-eating birds; on the other hand, the number of birds inhabiting the upper canopy of the forest declined sharply. All this unquestionably disrupted the normal food relations and ecological ties, and therefore extensive ecological work in such an *extremely* radioactive area would have limited scientific value.

In concluding this part of our discussion we cannot avoid noting the highly important fact that data on the study of radioecology in forest biocenoses in the Urals contaminated zone constituted the subject of a report included in the proceedings of the UN-sponsored Fourth Geneva Conference on peaceful uses of atomic energy, held in 1971. The report was published in Volume 11 of the proceedings, in 1972, but as is customary for UN conference reports from the USSR, it was published in Russian, and apparently for that reason no one paid attention to it or made it the subject of serious discussion. All three authors of this work (56) have already been encountered in previous sections of our review of this subject—Tikhomirov, Aleksakhin, and E. A. Fedorov. The report deals with the migration of radionuclides in forests and the effect of radiation on stands of forest trees. Judging by almost all the indications in the tables, the study was mainly concerned with the dynamics of strontium-90 *over a period of eleven years*, but the year the experiments began is not indicated. The report was given in 1971, but the factual material apparently reflects data from 1958 to 1969.

Materials for this type of international gathering are supposed to be presented almost a year in advance, and time was also required for the editing of the text once the research was completed and for the text to pass through the dual censorship required for publications going outside the USSR. The group of Soviet authors could not present a text lacking any methodology to an international conference with simultaneous translation into the UN's other official languages (English, French, and Spanish). Thus, for the "foreign audience," they prepared a brief description of how the contamination of the forest areas took place (the size of the areas was not given). Likewise, the calendar time of the contamination was not indicated, but since the observations on four species went on for eleven years, the initial contamination could have been in either 1957 or 1958.

The contamination of these forest areas came from above and in such cases it is customary to express the quantitative interaction between radioactive fallout and the trees in the form of a "retention coefficient," a quantity showing how much of the radioactivity settles in the canopy and lower leaves of the trees and how much reaches the ground.

For the rest, it seems appropriate to cite the work itself, with the table given by the authors (56, p. 678).

> The size of this coeffecient depends on the type and age of the stand of trees, the seasonal and meteorological conditions, and the physical and chemical form of the fallout nuclides. According to our observations, the retention coefficient varied according to concrete conditions with the following range of variations:

		(percent)
Young pine age 6–10 years, density of forest, 1.0	Spraying of crowns with Sr^{89} solutions	90–100
Stand of pine age 60 years, density of forest, 0.9	Fallout particles up to 50 microns	80–100
Stand of pine age 25 years, density of forest, 0.8	Fallout particles up to 100 microns	70–90

		(percent)
Stand of pine age 30 years, density of forest, 0.8	Fallout of secondary (soil) particles carried from the surface by the wind	40–60
Stand of birch age 40 years, before the opening of leaves, density of forest, 0.8	Fallout of secondary (soil) particles carried from the surface by the wind	20–25

Thus trees have a higher retentive capacity for radioactive fallout than do herbaceous plants (whose retention coefficient averages 25 percent). The factors contributing to this are the large biomass of the tree canopy, its greater differentiation, and the high ratio of surface to weight in leaves and needles, as a result of which the tree layer plays the role of a filter able to retain a considerable amount of radioactive dust. Since dry radioactive fallout settles to the earth's surface in quantities having relatively little weight, certainly not enough to saturate the canopy completely, we can apparently conclude that the retention coefficient of radioactive fallout by the tree layer is equal to the density of the forest with the exception of deciduous forests when the leaves are gone. The retentive capacity of the tree layer in that case is approximately 3 times less. (56, p. 678)

The first forest-contamination method listed was unquestionably experimental. The half-life of strontium-89 is 55 days, and the authors provide data on their observations indicating that the dispersal of the radioisotope was followed for 220 days.

The second and third contamination methods listed, "fallout particles" (which should be understood as particles containing radioactive isotopes, with no indication of their level) do not lend themselves to such a simple interpretation. The fact that the particles were of different sizes, "up to 50 microns" and "up to 100 microns," is not a rigorous experimental method. When carried by the wind, the larger particles (up to 100 microns) would reach the earth's surface before the smaller ones, and thus there would not be the even dispersion desirable in an experiment.

The fourth and fifth contamination techniques "fallout of secondary (soil) particles, carried from the surface by the wind," obviously could not be experimental. Wind erosion is seasonal, and dust storms do not occur every year. As the paper by Korsakov et al. (28) noted, the probability of radioactive dispersion with the spread of dust is greatest in spring and fall when the soil is bare, especially in *agricultural areas*. In the paper we are discussing, the wind erosion (which is possible only when there are strong winds) occurred in the spring, as is evident from the notation "before the opening of the leaves." If the dispersion of radioactive soil had occurred in the fall, the authors would have written "after the leaves had fallen." Of the five variants, the birch forest was the only one without leaves; and it cannot be compared with the evergreen pine forest. Under normal experimental conditions, a comparison would have been made between a birch forest with leaves and one without leaves. The total zone of wind erosion must have been *geographical*, because the pine forest was in podzol soil and the birch forest in a black earth region (chernozem soil). Both the forest and the wind erosion were real and fairly extensive, for the authors were able to observe an "edge effect."

"In the transfer of radioactive particles by the wind the particles are caught most effectively by the trees along the edge of a forest area. According to our observations, the quantity of radionuclides settling in the canopy along forest edges was 2–5 times greater than the fallout in adjacent nonforested areas: This 'edge effect' can be followed a distance of 15–20 meters in from the edge of the forest" (56, pp. 677–78). After the contamination of the forests by wind-borne radioactive soil, the observations were made for eleven years, as were the experiments involving the "fallout" of unspecified radioactive "particles." Thus the windstorm which contaminated the pine and birch forest with radioactive soil and the "original" contamination must have occurred at close intervals. To judge from the abstract of the report, the initial contamination included the radioisotopes strontium-90, cesium-137, zirconium-95, and cerium-144, but only the subsequent dynamics of strontium-90 in the forests was followed by the researchers.

The absolute contamination doses are not indicated in most of the experiments (after the first, with strontium-89 at a level of 1 millicurie per square meter). In the other experiments there is sometimes a reference to figures "adjusted" for millicuries per square meter, which implies an artificial mathematical averaging of the actual radioactivity. But the discussion of the "edge effect" has already indicated that the contamination was not uniform. Therefore, the tables on the migration of strontium by year essentially give relative magnitudes (percentages of the original contamination).

It is important to note, however, that in some areas the contamination level (of both birch and pine) was so high that it caused the trees to die out rapidly. Here again it is best to quote the authors themselves.

> A particular feature of radiation damage to a forest in the case of fallout is that when radionuclides occur, including fairly large-sized particles (up to 100 microns), the damage to leaves, needles, and tissue in the apical meristem is manifested above all in the upper and lower parts of the canopy, *especially on the windward side.* The topmost shoots maintain their viability while 95 percent of the canopy dies, and the apical meristem of the main trunk, although it is one of the most highly radiosensitive tree tissues, remains unharmed. This is explained by the relatively rapid cleansing of the topmost shoots through the action of wind and atmospheric precipitation. In contrast, when trees are subjected to external gamma radiation from a localized source the apical meristem reveals damage immediately and the drying out of the crown begins from the topmost shoots. (56, p. 683) [Emphasis added.]

The rapid lethal effect of certain fallout "particles," as here described, corresponds to the effect on a pine forest of chronic radiation at a level of 20–25 roentgens per day (7–8 roentgens per day of external radiation over five to six years being enough to kill pine trees [28]). From the excerpt just quoted it is evident that the "particles" in the original contamination were carried by the wind, not applied by artificial means (greater harm to the canopy is noted *from the windward side*).

An analysis of the other parts of this paper reveals that many more variants of the contamination were studied than the introductory part of the paper indicated. The authors seems to refer accidentally, in passing, to other types of contamination (or other levels) without citing any quantitative data. For example, they mention that birch trees were seven times more radioresistant than pine. This implies that birch-forest growth was also suppressed (by *secondary* fallout from wind-blown soil). Here the authors indicate that the difference between birch and pine may vary with the season (pine being twenty times more sensitive in the winter). It is evident from an accidental reference in another work by one of the authors (Tikhomirov, 50) that in some forest areas the concentration of strontium-90 in the soil reached 6–9 millicuries per square meter. This was equivalent to 10–15 rads per day, and after three years caused young trees to die. At a level of 3 millicuries per square meter the birch forest is damaged but survives (this level being more typical of many of Ilenko's experiments). The authors refer to Ilenko's work with mice, but it is not clear from the reference whether Ilenko did his experiments in the same birch forest. Nevertheless, it is quite clear from Tikhomirov's book on forest radioecology (50) that Ilenko's first publications did deal with the same ecological complex.

Despite the higher resistance of birch, the work by our authors refers to areas in which the radioactivity was almost completely lethal to birch stands as well as pine. This is indicated very briefly in the following paragraphs:

As an illustration of the combined operation of these factors let us examine the changes in the grass cover after the fallout of long-lived radionuclides over the forest. Where pine stands died out from radiation exposure and birch stands were severely damaged (30 percent of the trees drying up), the biomass of the grass cover increased by a factor of 3–5. At the same time, the species mix of the community was altered. According to E. G. Smirnov's data 2–3 years after the fallout the dominant species in the community came to be bushgrass (*Calamagrostis epigeios*)—a plant capable of vegetative reproduction through rhizomes protected by a layer of soil from beta radiation. This species was also fairly widespread under the canopy of the control stand of trees. With time such plants as *Chamaenerium angustifolium*, *Centaurea scabiosa*, and *Cirsium setasum* also became dominant. These are light-requiring, high-growing pioneer species with powerful root systems, and they force out even such radioresistant but relatively low-growing plants as *Eragaria vesca*, *Carex precox* and several others. (56, p. 685)

The capacity of stands of trees to revive after damage by ionizing radiation depends on the type of forest and the extent of the damage. In the conditions under which our experiments were conducted the deciduous stands—birch and aspen—maintained their capacity for revival through the formation of new shoots from roots or trunks, while the crowns dried up completely. Thus, in an area where as a result of radioactive fallout 70 percent of the upper part of young birch trees died out, along with about 30 percent in mature birch, after a year and a half there appeared an abundant growth from the trunks, which was fully viable and continued to develop for many years thereafter. When the damage is not total, trees revive after 8–10 years even if 95 percent of the crown has dried up. Since the topmost shoots in such a case remain unharmed as a rule, the trees did not appear outwardly different from normal ones after the crown had been restored. (56, p. 688)

Although the authors did not indicate the geographical location of their research, this paper (unlike others) did name the species of birch studied: *Betula verrucosa*. As we have said, more than fifty species of birch are found in the USSR, and *B. verrucosa* is typical precisely of the souther Urals, where it is found in the lower hills, other species occurring as one goes higher (54). We can make further deductions about the geographical location of this research by comparing the paper with the ones we have discussed above. For example, we know that one of the authors was E. A. Fedorov (who from 1964 to 1970 did joint research with A. I. Ilenko in the Chelyabinsk region). Fedorov was also a close associated of V. M. Klechkovsky, the man assigned in 1958 to organize the experimental station in the Chelyabinsk region. From 1951 to 1957 Evgeny Alekseevich Fedorov worked in the department of biochemistry and agrochemistry at the Timiriazev Agricultural Academy in Moscow, as I did. The department head at first was A. Shestakov, but after his death in 1954 it was V. M. Klechkovsky. A complete list of my own scientific writings includes two coauthored with E. A. Fedorov and published in 1956. As I have stated, the purpose of Kleckhovsky's special experimental station (which had a base in the Sverdlovsk region as well) was to study the effects of the contamination in the Chelyabinsk region on plants and the ecology in general.

It was natural for Klechkovsky to offer positions at the experimental station to his own young associates first of all. Some refused, as I did, because of the secrecy, but others accepted. One of them was E. A. Fedorov. After that he "disappeared" from science for many years. He produced nothing for "open" publication for ten years—until 1968, when his name suddenly reappeared, together with Klechkovsky's (57). But this publication was not a complete work; it consisted merely of the central abstract of a report by the two authors. Fedorov's first detailed paper (26)—written jointly with A. I. Ilenko—appeared only in

1970. The institutional affiliation listed, however, was Ilenko's. There were two Soviet reports at the 1971 Geneva conference. The second dealt with the distribution of radioisotopes in agricultural food chains, and among its several authors Fedorov is listed first. This time no institutional affiliations are listed at all, only the authors' country and the designations "Ministry of Higher Education" and "State Commission on Atomic Energy." There is no indication of the laboratory or institute with which any author was affiliated. As for the name of the experimental station, that remains a state secret, and I do not know it. Not concerned with classified information, I have analyzed only what was in the "open" press.

I was a close personal friend of Prof. Vsevolod Klechkovsky to the end of his life. He died in 1971 not long after his seventieth birthday. Prominent scientists are honored highly and with great ceremony in the USSR. Since Klechkovsky was an academician, he was given the major celebration due him on his seventieth birthday, with the presentation of an honorary order. At this ceremony I saw E. A. Fedorov but do not remember speaking with him. At the time I was a "dissident," who had just been dismissed from his job, and Fedorov held a high administrative post. Two years later I traveled from Obninsk to Moscow for the sad occasion of Klechkovsky's funeral. I regarded Klechkovsky as my second mentor in science, as I began my work during 1947 to 1950 in plant physiology under Prof. P. M. Zhukovsky. I do not remember whether I saw Fedorov at the funeral; there were a great many people and I might not have noticed him. But it is also possible he was not in Moscow that day. Obninsk is only one hundred kilometers from Moscow; the Urals are considerably farther.

From 1958 to 1971 Klechkovsky published dozens of scientific papers, but he placed his own name under only one item based on materials from the Urals contamination—the brief abstract we have cited (57). The rest of his publications involved research in Moscow that was truly experimental. In science,

Klechkovsky belonged to the school of Academician D. N. Prianishnikov, whose members were noted for their very conscientious and demanding attitude toward research methods and experimental details. To put his signature to a paper on an experiment whose methods and conditions were not clearly replicable was undoubtedly more than Klechkovsky would do.

At the 1971 Geneva conference dozens of questions were discussed. Volume 11, where the two reports by E. A. Fedorov et al. appear, contains material from several conference sessions. The reports by Fedorov et al. were part of the *radioecology session.* To judge from the text of Volume 11, questions were asked about the other reports and some sort of discussion was held. The texts of the discussions are printed after the reports. It is customary for reports in Russian to be accompanied by high-quality simultaneous translation. The report on forest radioecology which I have analyzed would inevitably have raised many questions on methodology and on general principles if it had been heard or read. It is surprising that not a single question was asked. In such cases there is a standard joke among Russian scientists: "So where was the chairman?" This time I would put the question differently: "Who was the chairman?" The answer can be found in the same volume, at the beginning of the section on ecology. The chairman of the radioecology session was—Sir John Hill.

Chapter 10

Field Plants in the Urals Radioactive Zone and Research in Plant Radiogenetics

In the contaminated zone, meadow plants grew not only in the usual meadow areas, but also, developing strongly, in the areas where forest stands had been destroyed by radiation. This kind of process, termed population replacement, is one of the subjects of radiogenetics. Thus, it is not surprising that much work was done with field plants in the radioactive zone by radiogeneticists. They were not interested in food chains or the particular species involved in the uptake of various isotopes from the soil. They were much more concerned with relative radiosensitivity, the dynamics of chromosomal aberrations, and the possible selection of more radioresistant forms.

The radiogeneticists who worked with plants were able to begin their research in this zone somewhat later than the zoologists. That is why publications of plant radiogenetics in the Urals biocenosis only began to appear in 1971. The main interest in this radioactive environment was shown by the Institute of General Genetics of the Soviet Academy of Sciences (director N. P. Dubinin) and the Institute of Cytology and Genetics of the Siberian Branch of the Soviet Academy of Sciences (director D. K. Beliaev). Researchers from the Institute of Forestry and the Forestry Laboratory of the Soviet Academy of Sciences and scientists from local experimental stations in the Urals (who usually did not list their institutional affiliations) also took a certain part in this work, as did researchers from the Institute of Agricultural Radiobiology, newly founded by V. M. Klechkovsky.

I have already referred to the extensive survey article by Dubinin et al. (49), which provides data on the cytogenetics of forest stands in the contaminated environment. This survey, published in 1972, gives a great deal of information on field plants, animals, and soil algae. But survey articles do not usually include information on methods; therefore, the best understanding of the essential research and the conditions in which it was done can be obtained from the original papers.

One of the first articles on plant radiogenetics in the Urals zone was published in 1971 in the journal *Genetika* (58). The article was submitted to the editors in the summer of 1970; thus it may be assumed that the research as such was completed in 1969; at any rate, the last samples for cytogenetic analysis could still have been taken in 1969. Although the work reflects many years of experiments, the authors do not give the actual dates of their research, and therefore I am obliged to make guesses. Similarly, the authors do not name the geographical location of their observations, although they indicate that the plants had existed in the radioactive environment for a long time. My conclusion as to the Urals location is drawn from sufficiently plain indirect evidence, as in so many other cases.

The purpose of the experiment was the usual one for almost all work in genetics. The authors wished to discover whether there were radiobiological adaptations—the appearance of radioresistant forms—as a result of the species' prolonged existence in a highly radioactive environment. Such changes could be the result of selection in populations, the development of more resistant somatic cells, or the "induction" of certain physiological-biochemical repair systems or similar adaptive mechanisms. The usual procedure for such experiments is to take seedlings from the seeds of species that have been growing in the radioactive environment for many generations, and other seedlings from control seeds (from "clean" plants of the same species) and subject both types of seedlings to external radiation under laboratory conditions. The expectation is that the seedlings from the radioactive environment will be more radioresistant and will display a smaller percentage of chromosomal aberration in the growth tissue.

In the article we are discussing (58), this experimental procedure was applied to four species of perrennial grasses. In this case the species names were indicated: *Agrimonia eupatoria*, *Libanotis sibirica* Rupr., *Centaurea scabiosa* L., and *Cirsium setosum* MB.

First of all, the research showed that when plant tissue is present for a long time in a radioactive environment it does become more resistant to external radiation. The species used point to a Urals location (a mixture of European and Siberian species on the same territory). And we can guess that the radioactive contamination was the same we know already as a product of the Urals disaster from the strontium levels indicated (1–3.7 millicuries per square meter) and from a statement already familiar to us from other such works: "The plants grew over a period of eleven years in areas whose higher radiation level was produced by the introduction, on one occasion, of the long-lived fission product strontium-90." A footnote mentions that, in addition to strontium, other radioactive products were present in the first application of isotopes to the soil (Zirconium-95, ruthenium-106, and cerium-144), but their activity was not followed since they have short half-lives. As we have noted, the gathering of material was evidently done in the summer of

1969; eleven years brings us back to the summer of 1958. This makes any date from the fall of 1957 to the spring of 1958 an appropriate one for the contamination of the surface. It can also be established easily that the authors were dealing with an extensive territory. First, all four species are cross-pollinating. Before the gathering of seeds was undertaken, there had to be certainty that *the plants had not been reached by "clean" pollen* from some neighboring area during the eleven years since the contamination. (Of course the authors do not mention *how* the territory was contaminated and do not date their own work from the time of the contamination.) Since pollen may be carried hundreds of meters or even several kilometers (in the case of wind pollination), there had to be certainty that *no seeds were carried in* from nearby "clean" zones (and the authors confirm that such was the case); consequently, *there were no "clean" areas in the vicinity.*

There are many fundamental methodological defects in the paper, which would not have been the case if the experiment had actually been planned in advance. There is no information on the total absorbed dose. It was impossible to estimate this because the authors did not know the original radioactivity of the isotope mixture. They indicate no variation in the strontium levels in the soil from year to year. It must have been reduced quite substantially over eleven years for many reasons. There were no cytogenetic measurements of chromosomal aberration immediately after the contamination, nor for the subsequent eleven years. It follows from the nature of the experiment that the authors gained access to the already existing radioactive biocenosis in 1967 or 1968, took seed samples, obtained old, approximate figures on soil radioactivity from the dosimetrists at the local experimental station (ignoring not only the short-lived isotopes but also the cesium-137, some of which remained even after eleven years), and carried out the rest of the work in the laboratory. These methodological shortcomings unquestionably reduce the scientific value of the work.

Two of the same group of researchers published a similar paper the same year, dealing with two species—cow vetch (*Vicia cracca* L.) and agrimony (*Agrimonia eupatoria* L.) (59). This was an entirely different environment, with each plant growing in an area with a different contamination level: the cow vetch in soil with 1.2 millicuries per square meter and the agrimony in soil with 1.5 millicuries per square meter. Both species are cross-pollinating and reproduce by seeds. It remains unclear why there were such different doses in the case of two comparable species. Besides this, the experiment suffered from the same inadequacies as the preceding one.

All these methodological defects are typical of a number of other papers in this series. There is no need to make a complete survey of them here.

I will cite only one of the last works in the series, published in 1975 (60). In the USSR the time between submission of an article to a journal and its appearance in print is fairly great. This article was submitted to the editors in June 1973; the seeds must then have been gathered no later than the summer of 1972.

The authors state that the experiments were done with plants *growing under radioactive environmental conditions for fourteen years.* In this case the authors have sharply increased the number of plant species, studying radioactive adaptation in eighteen species. The plants were taken from three different areas with strontium-90 concentrations of 0.3, 1.0, and 3.7 millicuries per square meter respectively. Two of these figures are completely consistent with the levels indicated in the work done several years earlier (58), although the strontium activity in the soil could not have remained constant in the very same area from 1969 to 1972. If the level of the *original* contamination from 1957 or 1958 *is meant, the soil radioactivity should have been remeasured, since the taking of seed samples began only in* 1972. In some soils as much as 6 percent of the strontium must have been washed away with the surface runoff during the course of one year. (These were podzol soils in a high-moisture area; see note 40, communication of E. B. Tiuriukanova.) But podzol soil has little calcium. In the southern Urals the calcium content, which fixes strontium-90 in the soil, is higher and the precipitation level is low (about 350–400 millimeters per year, nearly half the level in Byelorussia or the Moscow region). But here, too, radioisotopes are washed out of the soil and natural decay takes place, both processes necessarily becoming rather marked after fourteen years. Once again, the impression is created that the authors had no opportunity to make independent dosimetric measurements but were given cartographic information prepared long before 1972.

Chapter 11

Population Genetics Research in the Radioactive Environment

The wide areas contaminated by high doses of radioactivity from isotopes typical of the atomic industry constituted a unique opportunity for genetic studies of *populations*. Unfortunately this research was begun too late and without the necessary methodological arrangements for full cooperation between biologists and physicists. Much of the data on the nature of the contamination, the original composition of the radioactive mixture, regular dosimetric measurements, and certain other types of information have been kept classified thus far and were not available to biologists and ecologists. The experiments with plants discussed in the previous chapter could have had more value for population genetics if the authors had known, for example, how many generations of a particular species there were in this area over the eleven- or fourteen-year period. In work with perennials, one cannot simply state the year of the research, because the seeds gathered in any year will inevitably constitute a mixture from various generations. In work with animals, it is easier to calculate the number of generations; in the case of mammals, fencing must be erected to limit migration and restrict a substantial group of animals to a particular area with a particular level of radioactivity. There are significantly fewer methodological difficulties in the study of soil animals, because they do not of course migrate great distances.

In general, very little has been done on the radiogenetics of populations in the Urals contaminated zone. The survey article by Dubinin et al. (49) reported a population genetics study of two species of mice. No separate paper directly describing the experimental work that produced this data has been published; therefore the methodological details remain unclear. In addition, population genetics studies on the soil alga Chlorella were published for several years (61, 62, and 49).

N. P. Dubinin is the director of the Institute of General Genetics, and his own experimental work is usually done on the fruit fly—*Drosophila*. Which of the five coauthors of the survey article (49) did the research on mice remains unclear. The research aim was essentially the same as in the work by Ilenko et al. (33) published two years later, in 1974. Dubinin's group collaborated closely with Ilenko's, and the description of the contamination in the two areas where mice were caught refers to Ilenko's 1967 articles (20, 21). Dubinin and his associates suggested that the selection of new strains of mice, more resistant to radioactivity, could occur after many years habitation in a radioactive environment.

This proposition was tested in two species of mice living in the radioactive zone: the red field mouse *Clethrionomus rutilus* and the woodmouse *Apodemus sylvaticus*. The authors do not indicate when the experiment began nor the year in which they did the research on the radioresistance of the mice and the dynamics of chromosomal aberrations in their cells (relative to a control group). They do state, however, that "there had been about twenty-five to thirty generations" (49, p. 194) in the contaminated territory at the time the experiments began. It is not hard to arrive at a figure of ten to eleven years for that many generations. As I noted in chapter 6, new litters of mice typically resettle at fairly large distances from their native nests. Therefore the researchers had to be certain that mice caught after thirty generations had actually come from populations inhabiting the same environment for ten to eleven years. Such certainty could exist only if the contamination had affected a large territory, with a radius of at least some 30 kilometers. Dubinin and his associates do not indicate the size of the contaminated territory, but in the tables showing the effect of the radioactive environment on the frequency of chromosomal aberrations, the strontium concentration levels of the two areas are given in *curies per square kilometer.* In almost all the research on plants and animals that we have examined, the environmental radioactity was given per square meter. This is the first time the scale of "curies per square kilometer" (49 p. 196) has been used. Apparently this was an oversight by the censor; after all, even specialized censors of scientific publications cannot catch every tiny detail, especially in statistics, tables, and units of measurement.

Three biocenoses were compared: a control; one with 1,000–1,500 curies per square kilometer; and one with 1,800–3,500 curies per square kilometer. Converted to meters, this is 1.–1.5 millicuries per square meter and 1.8–3.5 millicuries per square meter. Ilenko, who worked in the same area, gave practically the same figures (3.4 rather than 3.5), and it is possible Dubinin rounded off to get "3,500." But if the radioactivity is measured in thousands of curies and in terms of square kilometers, it is immediately obvious that this could not be an artificially created experimental field.

In such experiments, where animals have lived in a contaminated zone for ten or so years, the changes in radioactivity, as we have said, should be given year by year. But this possibility did not exist for Dubinin et al. The authors conclude from their series of experiments that the "mutation load" noticeably increased in mice inhabiting the radioactive environment for a prolonged period and therefore the frequency of somatic mutation was also higher. However, the cells of these mice were more radioresistant to additional doses. In contrast to similar experiments by Ilenko et al. (33), in which radiosensitivity was determined by the mortality rate from external gamma radiation, Dubinin et al. measured radiosensitivity by the increased frequency of chromosomal rearrangement resulting from additional quantities of strontium-90 injected into the mice.

Another series of population-genetics studies was done with the single-cell soil alga *Chlorella*. All the material in this series was published by V. A. Shevchenko and his associates from the Institute of General Genetics. This group had done earlier research on the genetics of this alga. (Shevchenko had published many purely laboratory experiments on *Chlorella* radiogenetics.) In 1970 he published an interesting study on *Chlorella* radiogenetics *under natural conditions* (61). Work with soil algae does not require large areas. It can be done in several square meters under natural conditions in soil *experimentally* contaminated with strontium and cesium. There would be no reason to assume that their research on *Chlorella* was done in the Chelyabinsk region except that Dubinin states in his autobiography (51) that his colleagues' experiments with *Chlorella* were carried out in the same contaminated area in which they had made observations on other species *for eleven years*. As Dubinin puts it, in this area "some of the species evolved more radioresistant forms. There new populations ceased to suffer from the effects of certain doses of radiation. The single-cell green soil alga *Chlorella* was one such species. However, it took five years for a new radioresistant *Chlorella* to emerge through mutation and selection. During that time 200 generations passed, all of them existing in a high radiation background" (51, p. 330).

Actually that is not quite the way things were. Shevchenko began to collect soil samples for the study of the *Chlorella only five years after the natural region was contaminated by radioactivity.* The last publication in the series (49)—the survey article whose authors included Shevchenko and Dubinin—reports that the *Chlorella* samples "were taken five, six, and eleven years after the radionuclides were applied. Moreover, there had been approximately four hundred generations of *Chlorella* by the last date of analysis of the natural material" (49, p. 182). We are not told the years in which the samples were taken, but the annual survey published in 1972 was "sent for printing on November 22, 1971," according to the publication data in the volume.

In the USSR such major annual volumes are prepared for publication no less than seven to eight months in advance; therefore the survey article was completed in early 1971 or late 1970. (The references listed in the articles include no works published after *1969*.) Thus, the reference to eleven years since the beginning of the experiments with *Chlorella* leads us back once again to 1957–1958 as the time of the Urals disaster.

It is quite obvious that if Shevchenko's group had contaminated the soil *experimentally* in 1958, the examination of the algae for radioresistance would not have begun after two hundred generations or five years. However, the chief peculiarity of this work is that Shevchenko et al. gathered their research material from areas of *exceptionally high radioactivity.* It is not surprising that the authors do not indicate changes in the activity levels year by year. But they do not give the contamination levels in millicuries or microcuries either. Nevertheless, the radioactivity of the soil had to be measured in order to segregate strains of algae in relation to soil samples. The radioactivity is given in counts (disintegrations) per kilogram of soil. In the experiment, besides the control, there were six variant samples with different contamination levels. In the first case, the figure given for the soil activity was "from 1.10^6 to 1.10^7" disintegrations per minute per kilogram, and in the last, "from 1.10^9 to 1.10^{10}." One microcurie is equal to 37,000 disintegrations per second (approximately 2.10^6 disintegrations per minute). A millicurie equals 2.10^9 disintegrations per minute. If the experimental area had 1–10 millicuries per kilogram of soil, the radioactivity must have approached the magnitude of one curie for every square meter of soil 10 cm. deep. Such a concentration in chronic form is lethal for all animals and higher plants. For all practical purposes, only unicellular algae, which are among the most radioresistant of living things, could have withstood such doses, and survived with no noticeable suppression of growth.

In these experiments three species were taken from the soil: *Chlorella vulgaris, Chl. terticola,* and *Chl. ellipsoidea*. About 80 percent of the algae taken belonged to the first species, and the bulk of the subsequent experiments were done with that species.

The method for determining the soil activity is not indicated (the type of counter used, etc.), and therefore it is hard to judge to what extent such figures as 10^9 or 10^{10} represent readings with full allowance made for the efficiency of the instrument. The algae grew in the surface layers, and the samples were taken from the topmost layer, 1.2–0.5 centimeters in depth. The soil samples for measuring the activity levels were taken from the same layer. But it is precisely in this layer that the most substantial changes in relative radioactivity would have occurred by five years after the fallout of radionuclides to the surface. This layer must have been far more radioactive five years before the measurements began. The absence of the extremely important original data shows once again that the authors did not arrange the experiment themselves but made use of existing contamination long after it had happened.

Since radioactivity level in this part of the contaminated zone was nearly one curie per square meter, it may be assumed that this sector was close to the epicenter of the initial contamination. Surface vegetation could hardly have survived in these areas. We can only picture it as bare soil with some green from algae growing where there was moisture.

When an industrial nuclear accident occurs it is precisely in such "bare" areas that the greatest danger of the spread of radioactivity through wind erosion arises; therefore, the heavily radioactive soil must either be plowed under very deeply or be removed for storage deep underground. Why there were still such highly radioactive areas on the surface five years after the contamination remains unclear.

Chapter 12

The CIA Documents on the Urals Nuclear Disaster

As I indicated at the beginning of this book, when I first reported on the Urals disaster in November 1976, my account was immediately dismissed as "science fiction" by Sir John Hill. In the United States, the journalists turned, not to the Atomic Energy Commission (AEC), but to the CIA for information, and comments by CIA analysts were published in many newspapers on November 10–11, 1976. According to the CIA's unnamed experts, a nuclear catastrophe had indeed occurred in the Urals in 1958, but it was a "reactor accident," they said, and the results were quickly cleaned up.

The brief and unattributed comments that were published were enough to indicate that the CIA had information about the Urals explosion which had not previously appeared in the press.

Not long before, the Freedom of Information Act had been passed in the United States. Under its provisions government agencies are obliged to provide individuals or organizations with any *nonclassified* or declassified information they request. The cost of copying the relevant documents, the expenses involved in deciding what can or cannot be declassified, and certain other costs, must be paid by the requesting party. "Customers" are supposed to indicate in advance the amount they are willing to pay.

The Natural Resources Defense Council, a U.S. environmental group, quickly took advantage of this act and applied to various agencies for information on accidents at Soviet atomic installations in the Urals region from 1957 to 1961. Copies of these requests were made available to me. For those interested in the form such a request might take I include one of these letters at the beginning of the documents section (Exhibit 1) at the end of this book. We can see from the date that it was mailed five days after the first commentaries by the CIA and before the information of Professor Tumerman became public.

After this letter and a number of others were sent to the CIA, the Department of the Air Force, and similar agencies, a lengthy correspondence ensued that is not worth going into here.

Only in early February 1977 did the Natural Resources Defense Council finally receive a fairly detailed response, along with a number of documents. See the document section, Exhibit 2, for an excerpt from the CIA letter dated February 4, 1977.

The NRDC did not find these documents particularly informative, and the press did not report them at the time. Three other documents were obtained from other sources, but the NRDC did not find these very enlightening either. When they sent copies of the declassified documents to me, however, I discovered a great deal of useful information in them. Since there was no press coverage of this first declassification of CIA documents on the Urals disaster (only in late November 1977 did stories appear on the front pages of American newspapers, when the CIA declassified a number of additional documents), I consider it necessary to discuss these at some length. The later ones did not add much, and in most cases were misinterpreted.

In 1957, 1958, and 1959, as we know, U-2 spy planes were flying over the atomic centers in the Urals, continually photographing the region. Later this type of territorial intelligence was done by specially equipped artificial satellites. Obviously a vast amount of information can be obtained through this method. In addition, American intelligence agencies classify and analyze the reports of agents and defectors, and this abundance of data is apparently processed in all sorts of ways. Therefore it was hard to imagine that American and other Western intelligence did not know about, and had not investigated, the disaster which was known, if only in general outline, to millions of inhabitants of the central and southern Urals.

But just because Western intelligence people knew about it does not mean they would release their information. In fact, they kept it secret for quite a few years. Intelligence agencies are not quick to acquaint the press with the information they receive through diverse special channels. Not only that; the Urals disaster could not have been a desirable subject for possibly sensational coverage in late 1957 or early 1958. At that time there had been many news stories in the United States about the near-disaster at the Enrico Fermi reactor near Detroit, publicity which the government and the AEC found extremely unpleasant. In October 1957 a reactor accident occurred at Windscale in England. Antinuclear demonstrations and campaigns developed on a large scale in the United States and Europe. Under such conditions to "reveal" the nuclear disaster in the USSR would have had negative repercussions, not only (and not even mainly) against the Soviet government. Thus there was no reason to expect Western intelligence services to be quick about making known the tragic events at the center of the nuclear industry in the USSR.

But this is obvious to me only now. In 1976, when I first wrote on the subject in the *New Scientist*, I did not go into any detail because I assumed this event of many years vintage had ceased to be sensational. It proved otherwise.

I should comment on another one of the items the NRDC received. It was the text of a discussion in the CIA on an article in the *Washington Post* November 17, 1976, where the suggestion was made that the disaster I described had resulted from an earthquake recorded in the Lake Baikal region at the end of 1958. This was considered an unlikely explanation, since Lake Baikal is too far from the Chelyabinsk region (over 2,000 kilometers). According to some rumors (the source was not named), there had been another major accident in 1958 at Troitsk, a city south of Chelyabinsk, where some other nuclear reactors were apparently located.

Among the comments by CIA personnel on the *Washington Post* article, however, was a very interesting description of how the reactor waste was stored in the Chelyabinsk region, based on information from an unknown "source" (see Exhibit 3 in the documents section).

Included in the set of documents was an obviously important report, not from the CIA but the AEC. In the margin of the report are the words "classification canceled with deletions," an illegible signature, and the date December 23, 1976. I would have liked to begin with this document, because the original date on it is 1958. (Other documents contained information that became known later.) However, this document, AECIR 4-61, has been so well "sanitized" that everything before the editor's note on page 46 has been deleted. Pages 47 and 49 were also deleted in full and, again, only an editor's note has survived—on page 48. I include pages 46 and 48 to illustrate the appearance of a thoroughly "sanitized" document (see Exhibits 4 and 5 in the documents section).

The Urals disaster was apparently not the subject of the document as a whole, which was a secret report on the second

Geneva conference on the peaceful uses of atomic energy, held in 1958. The first editor's note declassified by the CIA, incidentally, could serve as an illustration to our chapter on the contamination of lakes and fish.

The next document, in the chronological order in which the CIA obtained its information, dates from early 1959, and the date of its distribution within the CIA was March 4, 1959. The document, in full, is shown as Exhibit 6 in the documents section.

Again, the source of the information remains classified and only a few sentences from the text are considered declassifiable. I should comment that Kasli is located near Kyshtym and Kamensk-Uralsky on the Sverdlovsk-Chelyabinsk highway, in Sverdlovsk province (Oblast). It is a large city with a population of over 150,000 (apparently about 100,000 in 1957–58). Professor Tumerman had to pass through this city in the trip he described (see Chapter 2).

The panic that developed in Kamensk-Uralsky was typical of other cities in the central and southern Urals. I was told of similar panics in Sverdlovsk and Chelyabinsk, where the sale of foodstuffs at private markets and collective-farm markets was also banned. Food was distributed only through the system of state stores during the first weeks after the disaster.

The next two "declassified" documents are of little interest. They simply report that there are many nuclear installations and reactors in the Chelyabinsk region. This is well known even without the CIA. Of considerably greater interest is the document distributed by the CIA (to those supposed to know) on February 16, 1961, and declassified on January 14, 1977. The document is reproduced as Exhibits 7 and 8 in the documents section. Points 1 and 2 were not declassified. The Techa River referred to in the document runs past Kyshtym and comes from Lake Kiziltash, a large lake of over fifty square kilometers.

Next comes a document received from Germany in December

1963. It reached the U.S. Air Force command in Europe on January 8, 1964. The declassified excerpts from the document are shown as Exhibit 9 in the documents section.

None of this information, although it undoubtedly came from people who lived in the Chelyabinsk region, bears the imprint of real knowledge of what happened. The disagreement on dates is typical. People were going by memory, and disparities on dates were inevitable. The assertion that it was an explosion involving "radioactive waste from a nuclear plant" tells us much more. To me, the very fact that the CIA had such information on the Kyshtym disaster is most important of all. Four documents mentioned in the CIA letter were in the possession of the AEC (later reorganized as the Energy Resource and Development Administration), and the CIA only noted the degree of classification of these documents. The two which were not declassified (April 1962 and April 1963) may have been more technical and specialized, so that their contents might have revealed the nature of the "source."

I assumed that after publication of my second *New Scientist* article, on June 30, 1977, containing many details on the radioecology of the contaminated region, I too could make use of the Freedom of Information Act to request data in the possession of the CIA about specific events in a particular area. I received a fairly quick reply to my letter to the CIA (see Exhibit 10).

At the end of November I took part in the 30th conference of the American Gerontological Society, held in San Francisco. On my way back to London I stopped in New York for a few days. While reading through the newspapers on Saturday, November 26, I came across an item on page 9 of the *New York Times* (Exhibit 11).

The same news appeared in the *Washington Post*, but in more detail, with the story beginning on page 1, column 1. As it turned out, both the *Washington Post* and Ralph Nader had called for declassification of the documents. Thus, the editors had obtained actual copies of the CIA documents, rather than simply cover the story with a "special" report from Washington. But even the *Washington Post* repeated the peculiar story of two explosions rather than one, the second being an atmospheric test over the contaminated Kyshtym area. Some of the new details were given in the *Washington Post* (Exhibit 12).

Other American newspapers reported approximately the same story. It was obvious to me that the possibility of a controlled atomic weapons test over the Kyshtym site was *absolutely* ruled out. This site was already seriously contaminated and therefore to build a mock village to test the effect of radiation on sheep and goats would have been pointless. The region between Sverdlovsk and Chelyabinsk is an industrial one and could not have been chosen intentionally as an atomic test site. One thing about these fourteen "declassified" documents was rather clear, however—newspapers are often in a great hurry to report sensational news without waiting to consult with knowledgeable experts.

After returning to London I found that CIA officer Wilson had kept his promise. The mail awaiting me included a letter from the CIA with the same set of documents that had been sent to the *Washington Post* and Ralph Nader. The documents had been sent to me a little earlier; the letter was dated November 14, 1977.

Of these fourteen documents, the first three were "complete and unabridged," but of no interest. The first two constituted a complete text of my own 1976 article in the *New Scientist*, which had been divided into two parts for broadcasting by radio to the United States from England. It found its way into the CIA file via the "Foreign Broadcast Information Service." The third "document" was a xerox of an article from the *Christian Science Monitor* of January 12, 1977, which discussed my 1976 revelations in connection with the testimony of Professor Tumerman. The remaining eleven documents were "sanitized." Documents 4 and 5 were already known from the previous release of documents (dated March 4, 1959, and February 16, 1961). Document 6 was simply a brief CIA memo of December 27, 1976, summarizing a discussion of Professor Tumerman's information. Document 7, dated May 21, 1958, was the one quoted by the *New York Times*. This was an analytical summary concerning conversations involving members of the Soviet pavillion at the World's Fair in Brussels in 1958. Among the exhibits at that fair there had been models of the first atomic power plant and other atomic energy installaitons. Apparently the conversations of personnel who explained these exhibits were overheard. However, the *New York Times* did not include the information that the accident was rumored to have happened in *the Chelyabinsk region.*

Document 8 (of the fourteen sent by officer Wilson), dated August 5, 1959, was new. But only the first two sentences of the entire document, not very big to begin with (only one page), had been "declassified" (see Exhibit 13).

Document 9, dated December 5, 1961, and document 10, dated December 21, 1951, are related, the second having the notation "supplement to document 3.202.034. of December 5, 1961." They relate the story of a mysterious explosion in the Chelyabinsk region (near the city of Emanzhelinsk) which caused many casualties. This is the report about hospitals filled with victims and the "horrible sight" of their skin sloughing off mentioned by both the *New York Times* and *Washington Post*.

To judge from the text, which I will not include here, the documents come from a second-hand source, a man who initially spoke with a woman about an "explosion" which this woman had not seen first-hand. At one point the "explosion" is described as having taken place in May 1960 and at another, in May 1961. The leaves of trees (poplars are mentioned) were covered with a red dust and soon curled up and fell off. The woman who gave this account said that the explosion had been approximately 50–60 kilometers from Chelyabinsk but she did not know in what direction. It is evident from paragraph 7 that the woman only knew about the "explosion" from an account by a man whose name is deleted in the CIA report.

The hospital with victims of the explosion was in Chelyabinsk itself. The description of this hospital was given by the woman from her own first-hand observation; she had been in the hospital, with the person who had told her about the explosion.

The site of the explosion is thought to be Emanzhelinsk in these two documents. This is a small city southeast of Chelyabinsk, not northwest of it, as it should be if Kyshtym was being indicated. Therefore this was either another accident altogether or the woman did not know the actual location (or date). Generally speaking, poplars are not typical of the Chelyabinsk region.

The man who passed on this information to the CIA (where it was recorded on December 5, 1961) took an interest in the woman's account, but since it was not very reliable, he later met and talked with the woman's acquaintance who had been the source of the "Emanzhelinsk" account. In document 10 the new "source" is obviously replying to prepared questions. From the repetition in his account it is clear that practically all of the woman's information came from him, except for the description of the hospital, where the woman had spent about three months (roughly May to July 1960). The hospital is said to be in the southern part of Chelyabinsk city. The "source" gives its address as "Green Store Street" [ulitsa Zelenogo magazina]—a very strange name for a street.

From all of this testimony it is evident that Emanzhelinsk was named in error. Neither the first nor the second "source" knew the actual site of the explosion, or even the approximate location, other than that it was a "forbidden zone." They are not at all certain about the date either. The only thing they are certain of is the presence of "explosion" victims at the hospital, because the woman had been in the same hospital as the victims, though in a different part, and the man had visited her there. According to her testimony, as given in English in document 10:

> The victims of the blast were placed in one wing of the hospital. None of them were permitted to leave this wing or to talk with other patients. Other patients were not permitted to talk with these victims or even visit them. Those who promenaded around the hospital grounds were all by themselves and the area was sectioned off so no one could get near them.

As document 11 (see Exhibit 14) the CIA offered the press No. OOK 323/20537-76. This document consists of two pages, the first of which is entirely "sanitized" with the exception of the heading. The document comes from the "foreign division of the CIA" and the "directorate of operations." The document reached the CIA only on September 20, 1976, that is, a month and a half before my first article in the *New Scientist*. It might be thought that someone from the editorial board passed on a preliminary text of my article to the CIA, but that variant is excluded. First of all, I wrote the article in late September and submitted it to the editors in early October 1976. Second, as may be seen from the only paragraph declassified for the press (there were five paragraphs in all), the "source" of the information speaks clearly of Kyshtym, and gives the "name" of the production site as "Chelyabinsk-40," an address we have already encountered in the declassified documents. But in my first article, I said that the accident occurred near the city of Blagoveshchensk. Blagoveshchensk is located approximately 150 kilometers southwest of Kyshtym. I remembered the name of this city from a conversation with V. M. Klechkovsky. He never used the word Kyshtym. This geographical distortion is also standard procedure. People from that "other world"—the circles where state secrets are discussed—are obliged to use geographic distortion in discussing events or facilities with people of "this" world, people without security clearances. But there was no great error in my article because in terms of the Urals and Siberia 150 kilometers really is not far.

Document 12, which all newspapers quoted as the most important, was a word-for-word recounting of Professor Tumerman's letter to the *Jeruselem Post* and of his subsequent remarks for other newspapers, radio, and TV. Someone in the CIA had gathered this material from ordinary newspaper sources (the wording coincides completely) and then made a summary document dated March 25, 1977. As for the 20-megaton atomic explosion which Soviet scientists allegedly set off in 1959 over the already contaminated Kyshtym region—a story which almost all the newspapers carried and which was in the CIA summary—it is described in document 13 (Exhibit 15), dated January 24, 1977. This document reached the CIA after my first article and after Professor Tumerman's account.

An unnamed "source" states in his report to the CIA that "it is possible" that the explosion in the southern Urals region reported in the press was in fact a nuclear weapons test. I include this last declassified document (Exhibit 16) in full simply to show that it has nothing to do with what is described in this book. It was an ordinary nuclear weapons test. There were dozens of such tests in the USSR and the United States from 1958 to 1959. If the region was the Urals (which is quite unlikely), we can only suppose that it was in the arctic zone of the Urals, where there is no population. In addition there is the large well-known arctic island of Novaya Zemlya (six to seven hundred kilometers long), which is used as a Soviet atomic testing site and which is, geographically, a continuation of the Urals range. In all probability the test explosion occurred precisely in this part of the "Urals," located two thousand kilometers north of Chelyabinsk.

Among the documents which are still considered classified, the CIA lists fifteen, documents 15 through 29. With the numbers, the dates are given and the names of the CIA officers responsible for their classification. I and other recipients of declassified documents were offered the right to request other, still classified documents if the reasons for such a request were explained. I have not yet exercised this right.

The earliest of the still secret documents is dated December

1958. Two from 1959 are classified. Eight documents reached the CIA from 1961 to 1962; and they have the initials "C.S.," which also appear on documents 4, 5, 7, and 8 as well. I would decode these as *C.* for Chelyabinsk and *S.* for Sverdlovsk. The coded initials "SD-KH" on one of the documents could likewise be decoded as *S* for Sver-, *D* for -dlovsk, *K* for Kys- and *H* for -htym.

Attached to the first thirteen documents was a CIA summary compiled as late as 1977. (Although the date is not given, the summary takes into account the report on the nuclear weapons test which reached the CIA in 1977.) It is quite obvious that this summary drew not only on "open" sources but also on those that are still "closed." If we compare this document with the known facts, we may conclude that the workers' settlement with the underground structures was the first atomic center in the Urals, site of the first military reactor to produce plutonium and the first reprocessing plant for plutonium separation. Two biographies of Academician Igor Kurchatov, scientific director and chief administrator of the Soviet atomic project (67, 68), report that construction of the "production site" began in 1945 and of the military reactor in 1947.

Kurchatov's biographers do not of course report such details as the use of forced labor to build the new city, although this fact is known not only from Solzhenitsyn's *Gulag Archipelago,* vol. 2, but from many other studies of Stalin's camps. Virtually all of the main atomic-industry facilities were built by prisoners. The book by I. N. Golovanov gives information about the beginning of the atomic age in the USSR, although there, too, everything is described in broad outline only. The construction of the first military reactor is treated in the following poetic way:

> Far from Moscow, at a picturesque site, the construction of a city got underway, with auxiliary installations and chemical plants. In January 1947 Kurchatov sent his most trusted aides to lay out the building that would house the first industrial uranium-fueled reactor and to assist in its construction.

> In the fall of 1947, when the frosts came, Kurchatov traveled to the construction site together with B. L. Vannikov. A large city had already grown up, populated by thousands of workers, technicians, and engineers of various professions. It was more than ten kilometers from the city to the construction site. . . . The first reactor was fueled with all of the metal uranium existing in the country at that time.

According to this book. Kurchatov spent several years in the new city. The first reactor went into operation in 1948. It is quite obvious that the construction of the "Kyshtym" site and of the first big reactor coincided in time completely. Since there was only enough metal uranium in the country for the production of plutonium in one industrial reactor, it is quite obvious that the Kyshtym region (or more exactly the region east of Kyshtym) was the main center for the initial production of atomic weapons. Since there was also a chemical plant for extracting plutonium, the first batches of nuclear waste also began to accumulate here. What to do with the nuclear waste was not then the main problem. It was seen as something requiring a solution "as things went along." The first solutions to this problem were not very satisfactory in the United States either and subsequently led to many dangerous situations. One could not expect anything better from the USSR. The American "fish soup," as we shall see in the next chapter, was also technically very primitive and many years later an explosion of atomic waste was avoided only because of the extremely dry climate and the fact that the water level in the region of the first U.S. military reactor was far beneath the surface. The atomic industrial center in the Soviet Union was not located in a dry region; there were lakes and rivers nearby. Water is needed for the constant cooling of the reactors and the chemical reprocessing plants. That is why a region with many lakes was chosen. But where nuclear waste is going to be stored, any water must be very much under control.

Chapter 13

The Causes of the Urals Disaster: An Attempted Reconstruction of the 1957-58 Events

As I said in chapter 1, my report on the Urals disaster and the confirmation of that report by Professor Tumerman came as a surprise to many experts. However, some people had previous knowledge of the event, and in addition to denials, certain explanations and interpretations were published, including the version that it had been a reactor explosion. The London *Sunday Times* of December 12, 1976, ran an article by Bryan Silcock entitled "The Mystery of Kyshtym," which gave a summary of the account by myself and Tumerman and then presented a skeptical comment on the possibility of such an explosion by an expert named Watson Clelland, who reported that he had seen several storage sites for reactor waste in the Soviet Union. "However," wrote Silcock,

> since the *New Scientist* article confirmation of a sort has reached the West from different but reliable sources. Ac-

> cording to these reports wastes *were buried* as Dr. Medvedev described. Their heat turned groundwater into steam, producing an explosion, but on a far smaller scale than originally alleged. There were, the sources say, no casualties.
>
> Dr. Clelland agreed that some kind of minor explosion could have occurred in this way, "but it could not possibly have affected hundreds of square miles," he said.

I think the material in the preceding chapters leaves no doubt that the contaminated territory was indeed on the scale of hundreds of square miles.

In the event of such widespread and heavy contamination involving millions of curies and occurring in a heavily populated industrial region, human casualties could not be avoided; people in the area would inevitably be exposed to severe radiation. It is enough to look at the detailed map of the southern Urals, or more precisely, the area between Chelyabinsk and Sverdlovsk (on p. 72), to grasp the potential consequences of even a slight delay in evacuation of the population.

Who the "reliable sources" mentioned by the *Sunday Times* were I do not know and have not tried to find out, since apparently it is a matter of confidential information. However, since my first article in November 1976 I have had occasion not only to research the various ecological materials set forth above and to acquaint myself with some of the CIA's information. I have also made a general study of the problem of nuclear-waste storage and its history, as well as of the accidents that have occurred in the atomic industry. In addition some unexpected new sources of information on the results of the Urals disaster have recently appeared. All of this permits me to attempt a reconstruction of the event and its consequences.

But first let us review very briefly some of the well-known and rather well documented nuclear accidents that have happened in Western Europe and the United States.

Reactor Accidents

Nuclear reactor accidents and even the explosion of a reactor are entirely possible if there is operator error or a malfunction, or if a major earthquake occurs. A somewhat dramatized description of the near-explosion of a reactor in a Detroit suburb may be found in the book by Fuller (63). Whether all the details in that book are totally reliable has often been debated. However, even if one takes a more moderate view of this event, it is obvious that nuclear engineering has not yet eliminated the possibility of such disasters.

Among the reactor accidents described in the scientific literature the best known is the one at Windscale, England, in October 1957. This accident is cited in practically all publications on accidents in the nuclear industry. According to the account given in Soviet sources (for example, the book by Tikhomirov [50]), at the time of its occurrence "about 20,000 curies of iodine-131, 600 curies of cesium-137, 80 curies of strontium-89, and 2 curies of strontium-90 were released to the atmosphere. The radioactive cloud that was formed was tracked for a distance of several hundred kilometers and passed over many European countries. The contamination of the area around Windscale, caused by the fallout of iodine-131 from the radioactive cloud, exceeded the permissible limits many times over and special measures were required to reduce the exposure of the local population" (p. 10).

It goes without saying that this 1957 accident was discussed in the British press for a long time and with great care. In addition to general news coverage and precise examination of the causes of the accident in books that stress the danger of atomic energy, such as Patterson's (64), a detailed report by a special government commission appointed to study all aspects of the accident was presented to Parliament by the prime minister in November 1957 (75).

The accident at Windscale is clearly not analogous to the Urals disaster—neither in the total amount of radioactivity nor in the composition of the isotope mixture. In those parts of the contaminated territory where samples were taken to test the radiosensitivity of *Chlorella*—in those areas alone—there was much more strontium-90 than was released to the atmosphere during the Windscale fire.

The article by Sir Martin Ryle (65) gives a chronology of other nuclear accidents in various countries in 1969, 1970, 1972, 1974, and 1975, none of which were listed as disasters but which could have caused major catastrophies.

Nevertheless, none of the known accounts of reactor accidents provide us with analogues of what happened in the Chelyabinsk region if the results of those accidents are compared with the type of contamination existing in the Urals territory.

Consequences of Atomic and Hydrogen Bomb Tests

The radioisotope mixture resulting from an atomic bomb explosion corresponds more closely to the isotopic composition of the Urals contamination. It was natural for the population from the evacuated parts of the southern Urals to spread rumors of such an explosion, although there was none of the destruction characteristic of a bomb. In the absence of any official account of the disaster, the legend that an atomic bomb exploded was widely circulated as the most likely explanation. Rumor undoubtedly had it that the explosion was accidental, since the heavily populated industrial area where the Soviet nuclear industry was centered could not deliberately have been used for one of the many atmospheric tests that took place in 1958. Incidentally, 1957 and 1958 were record years for atmospheric weapons tests: seventy-nine atomic and thermonuclear bombs were exploded in the United States in those two years, thirty-nine in the USSR, and twelve in Great Britain (64).

At the test sites used by the United States, for example, the vicinity of the Marshall Islands, much ecological and dosimetric research has been done. Such work may have been done near test sites in the USSR as well, but no results have been openly published there. Underground testing began in 1951 in the United States and in 1961 in the USSR. The underground explosions were all carefully programmed, however, and there were no major surprises, although one serious accidental result from one of these explosions is known—when radioactive material was released to the surface by an underground test in Nevada on July 6, 1962. The 100-kiloton device, exploded at a depth of approximately 250 meters (635 feet), produced a surface crater 1200 feet in diameter. Ninety percent of the radioactive contaminants released fell within a radius of about 2500 feet, but the remaining 10 percent spread over an area of 5000 square miles (66). A variety of radioecological studies were subsequently done in this area of almost total desert with sparse vegetation and a limited number of animal species.

A similar discharge of radionuclides to the surface occurred in an underground test in the USSR at the Novaya Zemlya test grounds in the Arctic. The results of the explosion were carefully studied and the findings published with photographs of the dynamics of the explosion and crater formation. A small book with all this information was published in the USSR, where the materials on the explosion (and it is possible the discharge was not accidental but experimental) were carefully studied to evaluate the feasibility of using atomic explosions for peaceful purposes. The conclusions drawn by the experts were negative: although the initial cost of the explosion was considerably less than for chemical explosives, the ensuing contamination ruled out the use of atomic explosions. The cost of decontaminating the environment was far greater than the expense that equally powerful chemical explosives would entail.

I have seen this book in Russian and it is undoubtedly known to experts in other countries, but unfortunately I do not have a precise bibliographical reference. It was a relatively recent publication with an exact description of the site of the explosion and its consequences.

Accidents Involving the Storage of Radioactive Waste and Other Fission Products from Reactors and Plutonium-Separation Plants

We know from the limited data on the history of the Soviet nuclear program, found mainly in biographies of Academician Igor Vasilyevich Kurchatov, the scientific director and administrator of this program, who died in 1960 (67, 68), that the construction of the first Soviet atomic bomb depended on the production of plutonium. The first military reactors were built in the Urals to obtain the necessary plutonium. The usual operational cycle includes a certain length of time for the fission of uranium in the reactor to produce plutonium, which is then separated from the fuel remnants and other substances at special chemical plants. After plutonium separation, a complex and varied mixture of highly radioactive isotopes remains, creating many problems because safe and proper storage must be arranged to last for *hundreds of years*.

Millions of curies of such concentrated waste remain after each "unloading" of a plutonium-producing reactor. Many of these isotopes have relatively short half-lives (lasting minutes, hours, or days). After several years the most dangerous remaining isotopes are of course strontium-90 (with a half-life of twenty-eight years) and cesium-137 (thirty years). Gaseous fission products such as carbon-14 are usually discharged into the atmosphere throughout the cycle.

The waste materials of the nuclear industry initially existed in liquid form, as acid or neutralized solutions. At the present time there are methods for the solidification of the waste—converting it into solid blocks. In the 1940s and '50s (and continuing through the '60s) most of the waste had to be stored for a long period in liquid form by one means or another. Since the shipping of concentrated and highly active liquid waste over long distances involves great difficulty and danger and because the quantity of such waste is enormous, running to millions of liters, large containers (tanks) were built of steel or reinforced concrete to store it, usually underground.

In the United States in the area around the Hanford reservation, other liquid wastes with lower radioisotope concentrations were poured into special trenches in which only the side walls were concrete. The water table in this area is far below the surface (more than one hundred meters) and therefore the soil under the trenches was considered a sufficiently safe fixative for the radioactive substances, capable of isolating them from the environment for centuries.

It was at the Hanford nuclear center, the largest U.S. industrial complex for the production of plutonium in the 1950s and '60s, that the two most dangerous situations developed. These can serve as analogues that will help us understand the possible causes of the Urals disaster.

The first accident is well known as the Tank 106-T leak, in which 435,000 liters of concentrated liquid waste escaped from a storage tank. A detailed analysis of the causes and results of this accident may be found in several published sources (64, 69, and others), in reports at a special hearing before the U.S. Joint Congressional Committee on Atomic Energy (70), and in the report on the investigation of this accident by the U.S. Atomic Energy Commission (71).

The Hanford atomic center is located near the city of Richland, Washington, in a desert reservation of about 750 square miles. It was established during World War II for the production of plutonium. For twenty-five years there were nine reactors in operation there, and several tens of thousands of kilograms of plutonium were produced. Through this process more than 70 million gallons of concentrated liquid waste were accumulated (over 300 million liters). To reduce the volume of liquid waste the solutions were slowly evaporated. The solid mass of salts thus formed was transferred to steel containers for further storage. However, in 1972 about 40 million gallons of concentrated waste were still in liquid form.

In June 1973 it was noticed that one of the larger underground waste tanks (Tank 106-T) had developed a leak. How long the concentrated radioactive liquid had been escaping remained unknown. The liquid had penetrated the soil and already spread contamination quite widely and to a depth of over forty feet beneath the tank. Fortunately the radioactivity did not penetrate to the water table two hundred feet below and therefore was not carried to the nearby river. After the discovery of the leak the liquid remaining in the tank was pumped out, but an estimated 115,000 gallons had run off into the soil. The leaked waste contained 40,000 curies of cesium-137, 14,000 curies of strontium-90, and 4 curies of plutonium, as well as a small quantity of other isotopes. The investigation of the incident revealed that other tanks also had leaks and that over a long period (since 1958) no less than half a million curies had leaked into the soil under the tanks. In this zone there were 151 containers for the liquid waste, whose total volume in 1973 came to tens of millions of curies.

If such a leak had occurred in an area where the water table was closer to the surface, the ground water would have quickly spread the contamination to a vast area, and after a certain time it would have worked its way into the soil, the vegetation, and the lakes and ponds of the surrounding region. Dozens of kilometers of territory with such a high contamination level could then become the source of extensive secondary contamination through soil erosion, dust storms, and other causes. This kind of contamination in the atomic industrial centers in the southern Urals, where liquid waste was stored, is entirely possible. In fact it is not excluded that *several* such areas of contamination exist, not just one.

A process of this kind could certainly have taken place in the Chelyabinsk region in particular, for the first big reactors for plutonium were built there (a development I will discuss in more detail below). However, the disaster of late 1957 to early 1958 was more sudden and was given the name "the Kyshtym disaster." According to the contradictory accounts of the disaster gathered by Professor Tumerman and recorded in the CIA archives, and according to two statements recorded on a television program "World in Action" (November 7, 1977, to be discussed below), what happened in the Kyshtym area was an explosion of nuclear waste stored underground.

A near-disaster that seems by all indicatious to be analogous to the one in the Urals was barely averted at the Hanford reservation—not in the tanks with highly active liquid waste, but in one of the trenches where the less active waste was poured. The volume of low-level waste was much greater, and it would have been very costly and required very large tanks to store the perhaps one billion (thousand million) liters of this liquid. Therefore a simpler and cheaper means of storage was worked out. It was simply allowed to soak into the dry earth that constituted the floor of the trench. Since the water table was so deep, it was felt that the ground would provide sufficiently stable and permanent storage for all this radioactivity. However, processes taking place in the ground over a period of many years created an unexpected situation in one of the large trenches.

As may be seen from the isotopic composition of the high-level waste in the tanks, the amount of plutonium was rather small, reckoned in curies. But with a half-life of 240,000 years, even four curies of this manmade element is a rather serious quantity. Nevertheless, it is not enough to cause an explosion, especially because in this case the plutonium was dispersed through a large amount of soil. But where did the plutonium come from in the first place? The radioactive mixtures are supposed to be "waste" remaining *after* the separation of plutonium. The recovery of the plutonium was the main purpose of the entire production process, and the cost of producing one kilogram of plutonium is twenty to thirty thousand dollars in the United States. Nevertheless, the techniques of plutonium separation have not been perfected to the point of 100 percent efficiency. Thus, some of the dangerous element ends up in the wastes.

Plutonium was also present in the isotope mixture poured off in less concentrated form into the trenches. It was assumed that the plutonium, since it is a poorly soluble element, would precipitate out, be largely adsorbed by the soil beneath each trench, and become fixed in a nonsoluble state. And that is what happened. But it occurred in a much narrower layer of soil than expected. The adsorption of the radioactive mixture by the soil took place along the lines of column chromatography (an ion-exchange process). In 1943–1944 column chromatography was not yet known. Different substances were separated by the soil at different depths, depending on their properties and molecular weights. The plutonium was adsorbed and *accumulated* in a relatively thin, upper layer of the soil rather than spread evenly throughout the whole mass. Over a period of many years, approximately 100 kilograms of plutonium accumulated not far below the dirt floor of one of these trenches (Z-9). This was a quantity sufficient to produce nearly a hundred small-size atom bombs (each with a destructive force equal to the bombs dropped on Japan in 1945). The volume of soil containing plutonium at trench Z-9 was approximately 1,800 cubic feet.

According to an official report—the WASH-1520 report (71)—of the U.S. Atomic Energy Commission, which investigated this problem and recommended the removal of the plutonium-contaminated soil with special equipment, conditions could have affected this high concentration of plutonium in such a way as to trigger a chain reaction, resulting in an explosion. A chain reaction could have been set off if water soaked into the plutonium-rich soil. The rapid heating of the water could turn it to steam and the pressure of the steam could produce an explosion, discharging the radioactive soil to the surface. One of the members of the group which investigated trench Z-9 defined this possibility as a "mud-volcano type explosion."

For a detailed description of the circumstances that could lead to such an explosion, I prefer to quote directly at this point from the summary of the report submitted to Congress by the Atomic Energy Commission.

> This Environmental Statement was prepared in accordance with the National Environmental Policy Act and in support of the Atomic Energy Commission's proposal for legislative authorization and appropriations for the design, construction and operation of the Contaminated Soil Removal Facility at Richland, Washington.
>
> The U.S. Atomic Energy Commission plans to remove plutonium contaminated soil from the floor of an existing enclosed trench (Z-9) used between July 1955 and June 1962 as a subsurface disposal facility for plutonium contaminated liquids from the Plutonium Finishing Plant on the Hanford Reservation near Richland, Washington. It is estimated that the soil to be removed contains approximately 100 kilograms of plutonium in a volume of approximately 1800 cubic feet.
>
> Liquid wastes from the Plutonium Finishing Plant (PFP) have been discharged to subsurface disposal facilities (enclosed trenches) since startup of the facility approximately 22 years ago. Most of the plutonium contained in these liquid wastes is sorbed (retained) by the soil and held within a few feet vertically of the point of release. The enclosed trenches are located in a fenced off area well within the boundaries of the controlled Hanford Plant. Careful surveillance using test wells has allowed this practice to be followed safely for 22 years. Due to the quantity of plutonium contained in the soil of the Z-9 enclosed trench, special precautions and emergency plans are required for Z-9 which are not required for other enclosed trenches.

> It is proposed to construct facilities at Z-9 to permit excavation and packaging of contaminated soil, to add equipment to the existing Plutonium Finishing Plant to permit recovery of plutonium from the contaminated soil, and to construct an Underground Storage Vault fourteen feet wide by eight feet high by 400 feet long for interim storage of contaminated soil.
>
> Removal of the plutonium contaminated soil will eliminate the need for special precautions and emergency plans necessary to assure the safe storage of the plutonium in the enclosed trench. Due to the quantity of plutonium contained in the soil of Z-9 it is possible to conceive of conditions which could result in a nuclear chain reaction. These conditions would be the rearrangement of the contaminated soil, flooding on the enclosed trench following a record snowfall and rapid melting (Chinook), and failure to implement planned emergency actions (pumping of flood waters from adjacent terrain and addition of neutron absorbing materials to the enclosed trench). Even though the probability of all these occurrences happening in sequence is extremely remote, removal of the Pu contaminated soil will eliminate any possibility of such an event. (71)

The AEC estimated that the necessary mining equipment would cost about $2 million, and this sum was appropriated for the work. It was proposed that the plutonium be separated from the soil, since 100 kilograms had a commercial value of over $3 million. But this was never done, and the soil dug from trench Z-9 was simply stored again elsewhere.

Altogether there were eleven trenches of this type at Hanford, but no critical situation developed at the ten other trenches, because according to the experts' estimates, those trenches altogether contained only 200 kilograms of plutonium.

From the time this atomic complex was founded, several million gallons of liquid waste had been poured into these trenches each year.

Various publications give different prognoses as to the consequences if there had been an explosion at trench Z-9. A number of experts hold that a mud-volcano type of explosion would have released large amounts of radioactivity into the atmosphere (73, 64). But the British expert Sir John Hill, in reply to L. Tumerman's letter, claims that radioactivity could not be discharged high into the air and that the contamination could not be carried much beyond the bounds of the trench itself (74). There is no question, however, that the exact consequences would very much depend on the weather conditions at the time. A strong wind or sudden snowstorm could expand the scale of such an accident to the point of ecological disaster.

Some skeptics have commented that the strontium-90 content is too great in the figures I have given for the spread of radioactivity from the Kyshtym site. It was apparently on the order of tens of millions of curies. In the opinion of these skeptics, that much strontiun could not have been accumulated in the waste in that area. But this skepticism is totally misplaced. Strontium accounts for between 4 and 5.7 percent of the radioactive mixture in fresh reactor waste (and of course a much higher percentage in "old" waste). According to American sources, there were 114 million curies of strontium-90 at the Hanford storage sites in 1976, and at another center of the U.S. atomic industry, the Savannah River plant, there were 150 million curies! Both figures apply to sites where high-level waste is stored (79). The strontium and cesium together at Hanford amounted to 360 million curies. At the Savannah River plant the combined amount was only 210 million, apparently because cesium is also extracted at Savannah River during the plutonium-separation process. Thus, the accumulation of tens of millions of curies of strontium in highly concentrated form at Kyshtym is quite conceivable.

It is also possible to suggest that cesium-137, a gamma source with a long life span, had been isolated and removed from the

Kyshtym center for some purpose other than research or medical use. At the beginning of the nuclear race, when the Soviet Union was too far behind the United States in atomic weapons, potentially lethal radioactive substances might have been stored as additional strategic material. This probably explains why there was so much strontium-90 and so little cesium-137 in the contaminated area.

My correspondent J. E. S. Bradley thinks, however, that the separation of strontium-90 and cesium-137 could have happened as a result of an ion-exchange process in clay minerals. Drilling holes with clay crusts, into which the waste was probably poured, had been made independently of the atomic industry. The area is rich in mineral resources and had many drilling holes, made both before and during World War II.

How the Explosion of Stored Waste Might Have Occurred: An Attempt at a Reconstruction

In 1947, when construction of the first large military reactor for plutonium production in the USSR began near Kyshtym, the technology of plutonium separation had not yet been completely worked out. Several months earlier I. V. Kurchatov had tested the first small experimental reactor for plutonium production near Moscow, and while the big reactor was in operation, the techniques of plutonium separation were being developed using fission products from the Moscow-area reactor. The first methods worked out by G. N. Yakovlev were imperfect and did not extract all the plutonium. More complete extraction of plutonium was achieved in a method developed by B. A. Nikitin and A. P. Ratner at the Radium Institute. This method became the basis for plutonium separation on an industrial scale at an associated plant in the same area (67, p. 68). The basis of the method was to initially dissolve the fuel rods from the reactor (with the original uranium) in hydro-nitric acid (HNO_3). I do not know how completely the uranium was transformed into plutonium or all the details of the method, but I think the process of separating the plutonium-nitrate involved crystallization of salts. After crystallization, a fairly large amount of the plutonium solution remained in liquid form (the "mother liquor"), which could be reused. But crystallization of the same solution, with the addition of new batches of dissolved fuel rods, could not occur many times, because dozens of other substances were present in the mixture of reactor waste and too great a concentration of those would affect the purity of the extracted plutonium. (The most modern plutonium-extraction methods, now used in the United States, ensure separation of 99.5 percent of the plutonium and uranium, with only 0.5 percent going off in wastes [81].)

In 1947 and 1948 the urgency of the project was so great that there was absolutely no time to work out all the details of the technology. Enough pure plutonium for several bombs had to be obtained quickly. The requirement was that the first bomb be exploded before the official celebration of Stalin's seventieth birthday. Kurchatov's group successfully coped with this task, carrying out the first bomb test in September 1949.

But the methods for storing the waste from plutonium production were also being worked out as things went along. Large steel containers (tanks) or concrete trenches were the most likely means chosen. On the basis of several structures of this kind that I have seen near the nuclear-power and radiochemical centers at Obninsk (where semi-industrial institutes for testing small reactors began to be built in 1946—half-institutes, half-prisons, with much of the work being done by prisoners, according to the custom of the time), I can assume that large-scale waste-storage facilities were located in forested areas. The United States was already conducting aerial surveillance of the main Urals areas at that time, and large forests were considered sufficiently reliable as camouflage.

It is quite possible that until 1953–54 all the main facilities for atomic-bomb production and associated waste storage were concentrated at one site—the one east of Kyshtym. Khrushchev drastically dispersed nuclear weapons production and the testing facilities for bombs and warheads. He had a mania for "decentralization" and insisted that major strategic installations be relocated so that they were far apart from one another. The decentralization of the Urals nuclear installations may well date from this time and probably involved the transporting of high-level waste over great distances as well. Khrushchev likewise introduced the use of prefabricated reinforced-concrete structures instead of poured concrete forms. If this method was used in making nuclear waste storage tanks, it could have resulted in many leaks.

I do not rule out the possibility that plutonium may have precipitated out in the tanks or filtered through the concrete and accumulated under them, producing a situation analogous to that in trench Z-9 at Hanford. There is more snow in the Chelyabinsk region than at Hanford, and the water table is closer to the surface. Thus, what was only a remote possibility in one case could have become a reality in the other. If the release of radioactivity to the surface occurred in the winter, a snowstorm could have spread it over great distances. (This could have happened in the spring as well, through surface erosion before the appearance of foliage.) In the winter, moreover, the soil freezes to a depth of one-half meter in the Kyshtym area. To break through this frozen layer, more pressure would have been required, resulting in a bigger explosion.

In the discussion of my first *New Scientist* article, a British physics professor, J. H. Fremlin of the department of applied radioactivity at the University of Birmingham, stated that there could not have been an explosion because there was only *one* reactor in the USSR in 1958. This nonsense was printed in an article by Lloyd Timberlake, Reuters science editor, "Facts Still Scarce on Nuclear Disaster in Soviet Union," the *Christian Science Monitor* of January 12, 1977. In fact, even the official biographies of Kurchatov (67, 68) which tell so little, make it clear that by 1957 the USSR was already a nuclear exporter, having started up reactors in Romania, Czechoslovakia, East Germany, Poland, China, Hungary, and Bulgaria (68, p. 181).

A small atomic power plant (the "first in the world") began operation in Obninsk in 1954, and reactors were built near Leningrad, in Central Asia, and in Georgia. However, it seems that the only plutonium-producing plants (evidently there were two) were still in the Urals, and the spent fuel from various reactors had to be delivered to the Urals and kept there in one form or another. I do not exclude the possibility that the explosion occurred where this more dangerous fissile material, containing large quantities of plutonium, was stored. In the initial haste of the period from 1947 to 1949, this material was apparently handled in a less than careful manner. Later it became possible to store it for about a year, so that the short-lived isotopes would decay and the chemical operations necessary for more complete extraction of plutonium would be made easier. The work by Korsakov et al. (28) was modeled on an explosion of reactor waste with the spread of radioactivity over a large area (the authors asserting that in their experiments the concentration of the dispersed material was only one curie per square kilometer). The authors sought to create a precise analogue to an accident at a facility for processing long-lived fission products held for 200–350 days after removal from a reactor; that is, spent fuel.

In one of the CIA documents quoted in the previous chapter (the discussion of the hypothesis that an earthquake damaged containers with radioactive waste), a CIA expert commented that the question of how highly active waste is stored in the USSR remains a secret and has never been discussed by Soviet scientists at international or other conferences or in the press. The Russians, as he put it, were only willing to discuss the disposal of medium and low-level waste. (We should comment that special attention must be paid to less active wastes as well; they are by no means "harmless." It was, after all, low- and medium-level waste that was disposed of at Hanford in the trenches where hundreds of kilograms of plutonium accumulated. The selective adsorption of the isotope, in effect, turned some of this material back into high-level waste.)

However, in two recent articles the various methods used in the USSR to store and dispose of not only medium- and low-level wastes but high-level waste as well are described at some length by Boris Belitzky, a scientific correspondent for Moscow Radio. The first article (76) was published in February 1976; the second (77) appeared after my first article and the testimony of Professor Tumerman—in April 1977.

In commenting on these articles I certainly do not propose to pass judgment on how dangerous one or another waste-isolation method may be, especially if it is a *current* method (recently developed in the USSR or other countries), for example, the methods of solidification and bituminization of liquid waste. These methods are not used everywhere, and they do not apply to the disaster that occurred over twenty years ago. As for the classification of waste levels in the USSR, wastes with less than 10^{-5} curies per liter are considered low-level, those with 10^{-5}–1 curies per liter are medium, and high-level waste is anything above that.

Belitzky gives very general details on the disposal of high-level waste in his first article.

> One method of disposing of high-activity wastes in the Soviet Union is to bury them, in special containers, in deep shafts lined with stainless steel and surrounded by reinforced concrete. Special precautions are taken to ensure that the shafts are waterproof. (76, p. 435)

In describing the storage of low- and medium-level waste, Belitzky (77) reports that in the Ulyanovsk region, where nuclear power facilities are located, more than 700,000 tons of liquid waste, accumulated over a ten-year period, were pumped under pressure into deep geological boreholes to a depth of 1400 meters. The sides of these boreholes were lined with waterproof minerals to isolate them laterally and they were isolated from overlying water-bearing strata by layers of waterproof clay. No migration of radioisotopes from these geological reservoirs has been observed, according to Belitzky, but it is only low- and medium-level waste that is disposed of in this way.

All these methods must be regarded as modern ones, however. It is not excluded that, despite the danger of leakage, the possibility of an explosion no longer exists in such cases. But to return to the circumstances at the first military center of the Soviet atomic industry at Kyshtym, we can assume with sufficient certainty that during the first few years of plutonium production the spent fuel from reactors was not kept long and was not constantly cooled (sometimes for a year), as is done today. There was no time for that.

It is evident from Kurchatov's biography that the first military reactor was started up only at the beginning of 1948. Kurchatov arrived at the construction site in the *fall of 1947* (66) when the reactor was still being built. "Throughout the construction period, Kurchatov came to the site every day and attentively followed the course of the work. . . . He made on-the-spot decisions. . . . Some accidents could not be avoided."

There is a description of the discovery of some borium in the reactor building. Borium is an element that is not supposed to contaminate graphite. Checking showed that there was borium in the linoleum covering the floor of the building. The linoleum was thrown out. After the basic structure was completed, the construction of the graphite core began. Let us quote Kurchatov's biographer: "The placing of the graphite was finished. Now came the most crucial phase—loading the reactor with uranium fuel. At this point Kurchatov, by the power of example, persuaded Vannikov to begin lowering the uranium rods into the channels, having the physicists monitor the neutron background continuously in order to know at any moment whether the reactor was about to go critical" (66, p. 71).

The author notes that the start-up of the reactor was a complex task and "not everything went smoothly" (read: there were accidents). "And all the while the government was inquiring about the progress of the work."

Even if we assume there were triple shifts around the clock, it is obvious that several months had to elapse between the time Kurchatov arrived and the point when the reactor attained full capacity. Thus, it is most likely that the reactor was started up in the *spring of 1948*.

"As the reactor came closer to full capacity unexpected corrosion phenomena appeared, along with the swelling of the irradiated uranium and graphite and other previously unknown processes. All the metal uranium in the country was used to fuel the first reactor. Tremendous resources were spent on its construction. It depended on Kurchatov whether the country would obtain the necessary plutonium by the appointed time or would suffer a delay."

By then thousands of workers in various fields were mastering the new technological process for plutonium production and the separation of uranium-235. "At this highly complex time, in August 1948, Kurchatov became a member of the Soviet Communist Party."

No further significant dates are given by Kurchatov's biographer, except for the date of the first Soviet atomic test—September 1949.

If the reactor accumulated plutonium for seven to eight months (the usual cycle being a year), the refueling of the reactor and removal of the spent fuel rods with their precious plutonium content began at the very end of 1948 or in early 1949. Undoubtedly the two to three hundred days needed for the rods to cool could not be spared. Psychologically no one was willing to accept such a waiting period. It is clear from the timing that the plutonium separation operations began immediately after the "unloading" of the reactor. Since the rods had to be dissolved in hydro-nitric acid (HNO_3) before the plutonium could be extracted, millions of liters of highly active acid (which apparently was later neutralized) must have accumulated at the plant. This highly active and unquestionably hot liquid was poured off into containers somewhere fairly nearby. There could not yet have been any reliable means of shipping liquid waste. All of the problems of *plutonium-cycle* nuclear technology were just then coming into view. As production expanded (according to Kurchatov's biographers, the plan was to add more reactors as soon as the first big one was successfully started up), new reactors were built around the plutonium-producing chemical plant. The nuclear center that arose in the Urals must have been the same type of complex—several reactors around a reprocessing plant—that existed in the United States at Hanford. The USSR began to project a hydrogen bomb as early as 1949. The need for plutonium increased. During the first several years the plutonium-extraction cycle must therefore have continued to leave out a long period in which the spent fuel rods were cooled with water.

Modern high-level waste tanks are built with double or triple walls, between which water circulates, and are covered with another thick layer of concrete. Those in charge of the Soviet atomic project at Kyshtym (P.O. Box 40) from 1948 to 1950 were undoubtedly not prepared for such structures.

It is useless to make any absolute assertions about the mechanism that set off the explosion in late 1957 or early 1958. Let us, however, review the possibilities.

It could have been an explosion of the type that nearly happened at trench Z-9: thermal heating of selectively adsorbed residual plutonium, resulting from a chain reaction begun by the moderating action of water on the plutonium.

It could also have been an explosion in a tank that was insufficiently cooled (for example, one having only a single cooling system, which for some reason failed) or a tank that had no cooling system at all. After plutonium separation, the concentrated waste gives off a great quantity of heat, especially in the first year—60 kilowatts per ton in the first several months, 16 kilowatts per ton after a year, and more than 2 kilowatts per ton after ten years (78). This amount of heat could build up great pressure, with the consequent danger of an explosion.

There is another possibility. In the USSR, according to the data in Parker (80), "liquid wastes," i.e., high-level wastes, are pumped under pressure into "authorized" geological formations and because of the high operating pressure these injections are extremely dangerous. An explosion is possible. It is hard to say whether such a method existed in 1957–58. I doubt it, because if such an explosion had caused the Urals disaster, caution would have dictated the abandonment of the method, and it would not now be in use.

One of my correspondents, Dr. J. E. S. Bradley, has proposed one other hypothesis to explain the Urals disaster.

> Extensive deep drilling had been performed in that area of the Urals, e.g., around Kyshtym, and the processing solutions were disposed of underground in some way in this geologically very complex area (the Urals form something of a hinge between Europe and Asia). At some later date, the residual plutonium in the solutions became concentrated by selective adsorption (probably in clay strata) and, in the presence of the abundant water, constituted a critical assemlby, which exploded (perhaps rather slowly and by a self-maintaining mechanism involving the concentration of solutions by the heat of the previous reaction). The products and the associated high-level wastes were vented along with much steam through the extensively jointed rocks in the area.

Truly, scientific imagination (or a capacity for "science fiction," if you will) is needed to construct hypotheses about the exact causes of the explosion. At any rate, it will be needed until such time as a factual description is given by those who were directly in charge of the first Soviet atomic center. But the fact that the explosion actually occurred, causing a great many casualties and contaminating a vast territory, and that it resulted from the improper storage of reactor products—of that there can be no doubt.

The Human Toll

No exact figures or related information can yet be given on the human casualties of the Urals disaster. Even in the case of earthquakes in the USSR, the number of victims is never reported. This even applies to earthquakes that happened thirty years ago, for which the present government could in no way be blamed. The new edition of the *Great Soviet Encyclopedia* states in its article "Earthquakes" that the October 1948 quake in Ashkhabad ranks with the worst in human history. It indicates that the city, capital of the Turkmen Soviet Republic, was *totally* destroyed. The earthquake happened at 4:00 A.M. when everyone was asleep. The population of Ashkhabad is known to have been about 200,000 in 1948. It was about 170,000 in 1959. For all other major earthquakes (in Japan, China, the United States, Turkey, and elsewhere) the article gives the number of victims, but in the case of Ashkhabad, this remains a state secret. The number of victims of mining, railroad, highway, and air-travel accidents are also kept secret. Atomic accidents are not an exception.

In discussing the Urals disaster, we should keep in mind that it occurred in a heavily populated area and covered a large territory. The evacuation was carried out belatedly and affected many thousands. As for details on the medical handling of the disaster, I know of only two. Prof. G. D. Baisogolov, who worked in the Chelyabinsk region and in 1965 was appointed deputy director of the radiology institute at Obninsk, and A. I. Burnazian, the deputy minister of health, received the Lenin Prize for developing effective methods for the treatment of radiation sickness. This award was not announced in the press. Evidently there were other scientists and health personnel in the group to whom the award was collectively given. A deputy minister would not be granted the Lenin Prize for some small-scale medical operation. To speak of *radiation sickness* is unquestionably to imply *severe forms* of that illness. Milder forms often go undetected. The worst doses, internal or external, kill the victims immediately. The less obvious lethal effects go on for weeks, months, and years. They are passed on to subsequent generations. Only a statistical estimate is possible and even that may never be made. No one knows the percentage of chromosomal aberrations that occur in areas where the atomic industry is concentrated, because in the USSR research of this kind is not only kept secret; it is absolutely forbidden. Comparisons of cancer mortality rates by region are also kept secret, as are facts on relative mortality from all sorts of other causes.

Thus one must go by rumors and guesses—and of course exaggeration is possible. But if the true picture is hidden, even from specialists, how can those be blamed who seek to learn the truth from secondary and indirect evidence?

In addition to the testimony of the several CIA "sources" quoted above on the large number of victims of the Kyshtym explosion and the crowded hospitals in the Chelyabinsk and Sverdlovsk regions even a year or two after the disaster, additional independent testimony has recently come forth. Among recent Soviet immigrants in Israel, the British television company Granada, which produced a program on the Kyshtym explosion, managed to find two witnesses who had lived in the southern and central Urals. Their testimony, broadcast in English translation in November 1977, is as follows.

> In recent years a new source of information on Russia has been growing, in Israel. However among the thousands of Russian Jews who have been allowed to emigrate only a very few come from the Sverdlosk region. World In Action were able to track two of them to their new homes. Because they still have families in the USSR they were unwilling to be identified.

Eyewitness number one left Russia in the early 1970's: This is his statement.

I lived with my parents in a village called Kopaesk, outside Chelyabinsk in 1948. Many people began arriving in Chelyabinsk and Kopaesk to live after being exiled from Kyshtym. Then we began to hear rumours that Kyshtym was being depopulated because a secret military plant was being built near there. We learned that the plant was called Chelyabinsk 40.

In 1954 I went to study at the Sverdlovsk Institute of Technology. As often as I could, sometimes every weekend, I would travel from Sverdlovsk to Kopaesk to visit my parents. I travelled by bus, car or train along a route that took me through the area around Kyshtym. It was very green and fertile with many villages, perhaps a village every 20 or 30 kilometers.

Around the end of 1957, we began to hear rumours that a terrible accident had occured at Chelyabinsk 40, that there had been a terrible nuclear explosion, an accident caused by the storage of the radioactive waste from the plant. Soon after the routes between Sverdlovsk and Kopaesk were closed. I could not see my parents for about a year.

Also during that year, I spoke with friends who were doctors. I went once to a hospital in Sverdlovsk for the removal of a wart, and one of my friends, a doctor, told me that the entire hospital was crammed with victims of the Kyshtym catastrophe. He said that all the hospitals in the entire area were crammed full, not only in Sverdlovsk but also in Chelyabinsk. These are huge hospitals with many hundreds of beds. The doctors all told me that the victims were suffering from radioactive contamination. It was a tremendous number of people, I believe thousands. I was told that most of them died.

Witness number two moved to a rebuilt Kyshtym in 1967. Although most of the radioactivity had been blown to the east of the town, the inhabitants were still living with the after effects, ten years later.

Now a nurse in Israel, she recorded her evidence with an English speaking friend.

There were no signs of destruction but everything which they bought at the market place or even if they went to the woods to gather mushrooms, they had to measure with radiometers and they had little radiometers with them.

When "she" came there, "she" got pregnant there and the Doctors told her to get rid of the child because of the radiation, they were afraid that something might be something irregular and so she had to make an abortion.

TV Commentary

These witnesses mention one other visible after effect of the accident. Around the countryside there were fenced enclosures containing piles of topsoil. On these, commonplace plants were growing in distorted shapes and sizes. They were known locally as "the graveyards of the Earth".

The accuracy of these two pieces of testimony is confirmed in a number of ways. The information is not directly linked with the CIA. Yet here, too, the address of the atomic industrial center is indicated as "Chelyabinsk 40." In the Soviet Union a post-office box number is the usual way of designating the address of a secret facility. Even the nuclear power research institute at Obninsk was referred to by a postbox number until 1968, when the city of Obninsk began to exist officially. Before that the institute was referred to as "Maloyaroslavets 2" after the nearest town. The removal of the most highly contaminated layer of topsoil by bulldozer is a common decontamination procedure. The fact that these piles of topsoil had not been stored again elsewhere, but were simply fenced off, testifies to the enormous amount of this super-contaminated layer (apparently from the same areas in which observations on *Chlorella* were made).

Kopeisk (not "Kopaesk") is a small town about 15 kilometers east of Chelyabinsk. What one of the witnesses says, about the removal of people from Kyshtym because military plants were being established there, is quite believable. It was a top-priority project and living quarters had to be cleared for the huge staff of specialists and construction personnel. The relocation of ten "corrective labor camps" to this site did not eliminate the need for dwelling places for "free workers"; only the heaviest and most dangerous work could be done by prisoner labor. All thc planning, designing, testing, development of methods, and so forth required thousands of "free" specialists and experts. Census and encyclopedia data indicate that in virtually all the cities of the Urals the population grew by a factor of two or a little more from 1939 to 1958. From various editions of Soviet encyclopedias we learn that Kyshtym had 16,000 inhabitants in 1926, 38,000 in 1936, 32,000 in 1958, and 36,000 in 1970. For a Urals industrial city located in a truly picturesque area, a population decline from 38,000 in 1936 to 36,000 in 1970 is something unique. During the same time the population of Sverdlovsk increased from 390,000 to 1,025,000—more than doubled.

When the witnesses quoted above spoke of "villages" every 20 or 30 kilometers, there was an inaccuracy in translation. The Russian word was *posyolok*. For "village" Russian uses *derevnya*, meaning a small agricultural settlement, usually the center of a collective farm or state farm. A *posyolok* is a *worker's settlement*—a small town built around or near a factory. In the industrial Urals there are indeed small workers' settlements every 15 or 20 kilometers between Sverdlovsk and Chelyabinsk: Kasli, Novogorodny, Karabash, Kaslinskoye, and so on. Each has some fifteen to thirty thousand inhabitants. Before the area was contaminated, its total population seems to have been approximately two hundred thousand. The industrial towns have been decontaminated and covered with new layers of asphalt. The villages (and agricultural areas) remain desolated to this day.

I leave this book without any separate "conclusion." There is no doubt that the Urals disaster was the biggest nuclear tragedy in peacetime that the world has known. It produced the largest radioactively contaminated ecological zone in the world. It will not be gone even a hundred years from now. When people will reinhabit this region is hard to predict. But I hope a time will come when there will be no need to keep such secrets, and that monuments will be built near Kyshtym both to the prisoners who died building this military-industrial complex and to the later victims of the Kyshtym disaster. To have only one monument—the giant head of Igor Kurchatov, at Kurchatov Square in Moscow opposite the institute where the Kyshtym atomic complex was planned—is certainly not enough. Something else is needed to honor those who helped lay the foundations of the Soviet Union's nuclear might at Kyshtym—and to mark their tragedy.

A GLOSSARY OF CERTAIN TERMS NEEDED BY THE NON-SPECIALIST

A Glossary of Certain Terms Needed by the Non-Specialist

Beta Radiation: streams of beta particles emitted by radioactive isotopes. Among the most common isotopes giving off beta radiation are radioactive carbon (C^{14}), radioactive phosphorus (P^{32}), radioactive strontium (Sr^{90}), and radioactive sulfur (S^{35}). Beta rays are essentially electrons traveling at high velocity, but the energy of beta particles is not so high as to allow them to penetrate very far into solid material. Even relatively thin layers of protective material are sufficient to stop beta rays. Living tissue may be penetrated by beta particles for small distances, varying from several millimeters to 2–3 centimeters, rarely as far as 5–8 centimeters.

Gamma Radiation: shortwave electromagnetic radiation consisting of high-energy protons emitted at the speed of light by the nuclei of certain radioactive isotopes, in addition to the emission of beta or alpha particles. Because of the high energy involved, the exposure of living tissue to gamma rays causes severe harm to internal organs. Protection against gamma rays is provided by thick layers of concrete, lead, or other materials. Isotopes that give off gamma rays include cobalt-60 (Co^{60}), iodine-130 and -131 (J^{130}, J^{131}), iron-55 and -59 (Fe^{55}, Fe^{59}), and cesium-137 (Cs^{137}).

Radiation Dose: measured in roentgens. Even before manmade radioactivity, the degree of radiation exposure was calculated in terms of exposure to X-rays (roentgen rays), and therefore the unit of measurement was given the name *roentgen*. The "biological equivalent of a roentgen" is often abbreviated BER. Absorbed dosage is measured in *rads* (an acronym from the phrase "radiation absorbed dosage"). Different plants and animals have different degrees of sensitivity to radiation. A fatal dose for mammals and humans is from 500 to 600 roentgens, more commonly expressed as 500 to 600 rads. However, the sensitivity of different tissues varies, and a lethal dose is defined as one causing the death of bone-marrow cells. Other tissues can withstand higher doses when exposed locally. Plants and lesser animals can withstand several thousand roentgens; bacteria and algae, tens of thousands.

Permissible Dose: the amount of radiation—or the concentration of radioactive isotopes in an environment—that is assumed not to cause harmful effects. Permissible doses vary with the type of isotope (the type of radiation emitted), the conditions of the exposure, the length of exposure, and other factors.

Curie (Ci): unit of radioactivity, named after Marie Curie, who discovered radium. This unit is equal to the quantity of radioactivity given off by any radioisotope whose decay rate is 3.7×10^{10} disintegrations per second. Other units of radioactivity are named accordingly: megacurie (a million curies), kilocuries (a thousand curies), millicurie (mCi, a thousandth of a curie); and microcurie (μCi, a millionth of a curie). A microcurie is equivalent to 37,000 disintegrations per second. In experimental work with radioisotopes on plants and animals, the quantities involved are most frequently measured in microcuries. Internal application of radioactive phosphorus in amounts of 30–40 millicuries can, for example, prove fatal to human beings. Strontium-90, by contrast, can be dangerous at the level of 1–2 millicuries because it is not quickly eliminated from the organism, but lodges in the bones, producing "chronic exposure."

Half-Life: the time it takes for half the radioactive isotope in a given sample to disintegrate and correspondingly for the danger it represents to diminish by half. Isotopes with a short half-life, measured in seconds, hours, or days, are considered generally less dangerous to the environment (for example, radioactive iodine, phosphorus, and sulfur). Strontium-90 and cesium-137 are the most dangerous waste products of nuclear reactors and atomic explosions because these isotopes have a long half-life (about thirty years). Therefore, when the environment is contaminated by these isotopes, the danger can last hundreds of years (for many half-lives). Carbon-14 has a half-life of more than five thousand years, but since it is released in gaseous form as carbon dioxide, the danger from radioactive carbon is thought to be less.

Nuclear Waste: what takes place in nuclear reactors is the controlled decay (or fission) of a nuclear fuel (mainly uranium); this is accompanied by the accumulation of a large quantity of radioactive isotopes. A great amount of heat is given off during this process, which in power reactors is used to generate electricity. The process lasts many months, the length of the cycle varying with the "heat production rate" of the reactor; that is, how rapidly the fission is allowed to take place. After the nuclear fuel is "burned," the spent fuel rods contain many millions of curies of various isotopes. The isotopes of importance in the manufacture of nuclear weapons are uranium-235 (U^{235}) and plutonium (Pu). It is easier to extract plutonium, a synthetic element produced in reactors, from the mixture of isotopes in spent nuclear fuel than it is to obtain U^{235} from natural uranium ore. That is why plutonium is used in the military nuclear industry. Before the extensive construction of nuclear power plants, the military nuclear industry was primarily engaged in extracting plutonium from the decay products of reactors. In order to extract the plutonium, the spent fuel rods must first be dissolved in strong acid.

The plutonium separation process is performed at special radio-chemical installations called reprocessing plants. Other radioactive isotopes can also be separated out for use in medicine and for other scientific purposes (for example, cobalt, iodine, phosphorus, and cesium), but the need for these is minor.

Most of the remaining radioisotopes, therefore, constitute *nuclear waste,* which must be stored in such a way that for centuries it cannot do harm to the environment and to human beings.

If the only thing needed is the heat energy from the reactor, the process of storage is simplified, since the waste can be stored in the solid form of the original rods. If the waste must be reprocessed for plutonium production, the problem is made more complex, because what must be stored is *liquid waste* with radioisotopes in different degrees of concentration (there are high, medium, and low levels of concentration). And there are *millions of liters* of these radioactive solutions.

Storage of Radioactive Reactor Waste: there are many ways of "storing" radioactive waste. The difficulties involved in storing liquid waste with a low concentration of isotopes (low-level waste) is the enormous volume of such waste. With high-level waste the main problem is that the continuing spontaneous radioactive decay gives off so much heat that water temperature is raised above the boiling point. Therefore containers with concentrated liquid waste must be constantly cooled. After a year or two of such forced-cooling the short-lived isotopes decay and the waste, now containing mainly the long-lived isotopes (the chief "remnants" being strontium-90 and cesium-137), can be treated in the same way as medium-level waste. Because both reactors and high-level waste containers require constant cooling (usually by circulating water), the centers of nuclear industry are located near rivers or large lakes in sparsely populated regions. The water of the lakes and rivers is used to cool the nuclear installations. The containers with concentrated waste may be above or under the ground. Medium-level waste is usually stored underground, sometimes at a great depth. Low-level waste is often simply dumped into rivers and large lakes. In England, the dumping of low-level waste directly into the sea is permitted, since the reprocessing plant is located on the seacoast. As a result, the concentration of radioisotopes in the Irish Sea is several times greater than in the ocean. In the USSR, the concentration of radioisotopes at present in the Caspian Sea, which has no runoff, is 20 times greater than in the ocean, but thus far it is no higher than the "permissible" radiation dose.

DOCUMENTS SECTION

Although several of the following documents are nearly illegible, they are facsimiles of the originals as received by the author.

Documents Section

Freedom of Information Act Officer
U.S. Energy Research and Development
   Administration
Washington, D.C. 20545

RE: Freedom of Information Act Request

Dear Sir:

I hereby request the documents identified on the attached sheet. This request is made pursuant to the Freedom of Information Act, 5 U.S.C. § 552, and ERDA's regulations thereunder.

Pursuant to 10 C.F.R. § 709.6, I hereby state that I will pay promptly any costs incurred in this request up to a maximum of $25.00. However, I hereby request that any fees chargeable under section 709.12 be waived on the ground that disclosure of the documents requested primarily benefits the public interest.

Exhibit 1. Sample letter of request for information from the CIA.

"The Energy Resource and Development Administration, in responding to your Freedom of Information request, recovered the documents listed below which originated with this Agency. They have been forwarded to us for review and direct response to you.

We have reviewed the documents and find that the first two are releasable in sanitized form, but that the remaining two are not releasable. I have cited the applicable exemptions from the Act beside each of the items in question.

	Document	Exemption
1.	Report No. CS-3389,785, 4 March 1959.	(b)(1), (b)(3)
2.	Report No. CS-K-3/465,141, 16 February 1961.	(b)(1), (b)(3)
3.	Report No. 00-B-3/256,712, 5 April 1963.	(b)(1), (b)(3), (b)(6)
4.	Report No. CS-K-3/507,314, 16 April 1962.	(b)(1), (b)(3)

For your information, exemption (b)(1) of the FOIA applies to material which has properly been classified under Sections 1 and 5(B) of Executive Order 11652. Exemption (b)(3) is pursuant to the Director's responsibility, under the National Security Act of 1947 and the CIA Act of 1949, to protect intelligence sources and methods from unauthorized disclosure. Exemption (b)(6) applies to information from personnel, medical and similar files, whose disclosure would be an unwarranted invasion of another's privacy."

Exhibit 2. Excerpt from CIA letter dated February 4, 1977, in response to request for information made by the Natural Resources Defense Council, a U.S. environmental group.

---

8. "The steel storage tanks above and below ground were said to have been no farther apart than 20 feet, meaning that when the first tanks ruptured, they splattered pieces of steel, breaking open the others."

Comment: As far as is known, all Soviet waste storage tanks are buried at least at ground level and the space between them is backfilled with dirt, which would tend to inhibit explosive propagation from one tank to another, even assuming a sufficiently violent explosion could occur that would shatter a tank.

9. O'Toole: "At the time of the accident highly poisonous radioactive cobalt, barium, cesium and strontium had been in the storage tanks about 10 years, the sources said."

Exhibit 3. Description of reactor-waste storage, taken from CIA documents which were declassified (with deletions) and released in December, 1976.

~~SECRET-NOFORN~~

AECIR
4-61

deleted

(Editor's Note: Some confirmation of this report can be gleaned from a conversation *deleted* with a Soviet Scientist in the UN Scientific Secretariat for the 2nd Geneva Conference of 1958 when the scientist told of catching a delicious fish meal in a lake near where he worked (location unknown) and it was only after consuming the meal did he learn that the fish were contaminated with radioactivity. The Soviet scientist told *deleted* that the lake had since been cleaned up. *deleted*

- 46 -

Exhibit 4. A thoroughly "sanitized" document, taken from a lengthy report written by the Atomic Energy Commission on the second Geneva conference, the topic of which was 'peaceful uses of atomic energy.'

* AEC editor, not the editor of this book.

~~SECRET-NOFORN~~

AECIR
4-61

deleted

(Editor's Note: It is noted that US scientists gathered that the USSR had had a "Windscale type" accident in one of their research reactors whose actual location is unknown,

deleted

p. 49 deleted in entirety

- 48 -

~~SECRET-NOFORN~~

Exhibit 5. Another "sanitized" document. See Exhibit 4 for further explanation.

* AEC editor, not the editor of this book.

INFORMATION REPORT INFORMATION REPORT

CENTRAL INTELLIGENCE AGENCY

This material contains information affecting the National Defense of the United States within the meaning of the Espionage Laws, Title 18, U.S.C. Secs. 793 and 794, the transmission or revelation of which in any manner to an unauthorized person is prohibited by law.

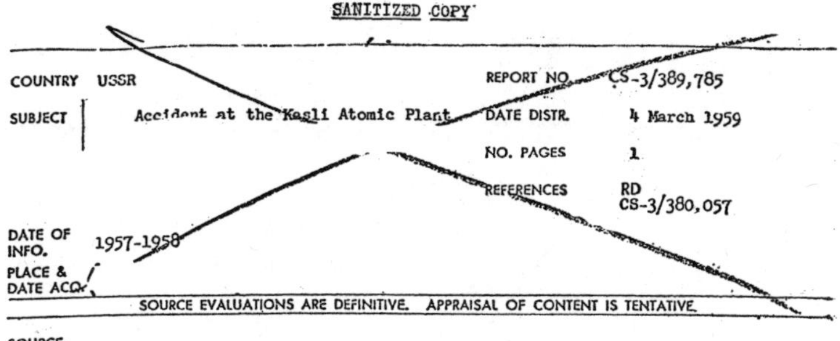

SANITIZED COPY

COUNTRY	USSR	REPORT NO.	CS-3/389,785
SUBJECT	Accident at the Kasli Atomic Plant	DATE DISTR.	4 March 1959
		NO. PAGES	1
		REFERENCES	RD CS-3/380,057
DATE OF INFO.	1957-1958		
PLACE & DATE ACQ.			

SOURCE EVALUATIONS ARE DEFINITIVE. APPRAISAL OF CONTENT IS TENTATIVE.

SOURCE:

In the winter of 1957, an unspecified accident occurred at the Kasli (N 55-54, E 60-48) atomic plant
All stores in Kamensk-Uralskiy which sold milk, meat, and other foodstuffs were closed as a precaution against radiation exposure, and new supplies were brought in two days later by train and truck. The food was sold directly from the vehicles, and the resulting queues were reminiscent of those during the worst shortages during World War II. The people in Kamensk-Uralskiy grew hysterical with fear, with an incidence of unknown "mysterious" diseases breaking out. A few leading citizens aroused the public anger by wearing small radiation counters which were not available to everyone.

Exhibit 6. "Sanitized" CIA document describing the town of Kamensky-Uralsky, a town near the Kalsi Atomic Plant, after an "unspecified" accident in the winter of 1957.

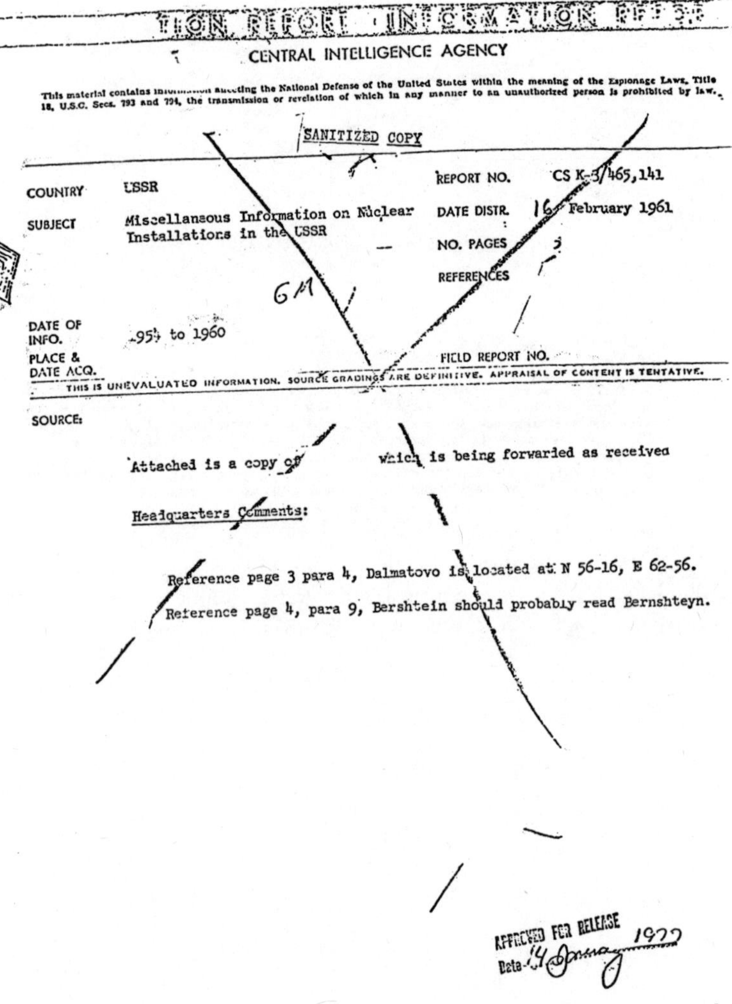
INFORMATION REPORT INFORMATION REPORT

CENTRAL INTELLIGENCE AGENCY

This material contains information affecting the National Defense of the United States within the meaning of the Espionage Laws, Title 18, U.S.C. Secs. 793 and 794, the transmission or revelation of which in any manner to an unauthorized person is prohibited by law.

SANITIZED COPY

COUNTRY	USSR	REPORT NO.	CS K-3/465,141
SUBJECT	Miscellaneous Information on Nuclear Installations in the USSR	DATE DISTR.	16 February 1961
		NO. PAGES	5
		REFERENCES	
DATE OF INFO.	1955 to 1960		
PLACE & DATE ACQ.		FIELD REPORT NO.	

THIS IS UNEVALUATED INFORMATION. SOURCE GRADINGS ARE DEFINITIVE. APPRAISAL OF CONTENT IS TENTATIVE.

SOURCE:

Attached is a copy of which is being forwarded as received

Headquarters Comments:

Reference page 3 para 4, Dalmatovo is located at N 56-16, E 62-56.

Reference page 4, para 9, Bershtein should probably read Bernshteyn.

APPROVED FOR RELEASE
Date 14 January 1977

Exhibit 7. Cover memo for CIA report written in February, 1961, but not released until January, 1977.

-3-

Kyshtym

3. In spring 1958, he heard from several people that large areas north of Chelyabinsk were contaminated by radioactive waste from a nuclear plant operating at an unknown site near Kyshtym, a town 70 kilometers northwest of Chelyabinsk on the Chelyabinsk-Sverdlovsk railroad line. It was general knowledge that the Chelyabinsk area had an abnormally high number of cancer cases. To go swimming in the numerous lakes and rivers in the vicinity was considered a health hazard by some people. Food brought by the peasants to the Chelyabinsk market (rynok) was checked by the municipal health authorities in a small house at the market entrance where the peasants also paid their sales tax. How radioactive food was destroyed was unknown to source. Food delivered to the plants, schools, etc., by the kolkhozy and sovkhozy was probably examined by the latter themselves. Until 1958 passengers were checked at the Kyshtym railway station, and nobody could enter the town without a special permit. By what authority the permit was issued and why the checking was discontinued in 1958, source was unable to say. In addition, some villages in the Kyshtym area had been contaminated and burned down, and the inhabitants moved into new ones built by the government. They were allowed to take with them only the clothes in which they were dressed.

4. The plant was probably processing radioactive deposits found in the Urals, among which were huge deposits of zirconium. Source was told this by a friend who, in 1953-1954, had a job in the Kyshtym-Argayash area. He also told source that as early as 1954 that the water of the Techa River, running from Lake Kyzyltash and Lake Ulagach and emptying into the Iset River at Dalmatovo, had become highly radioactive.

5. In late August 1960, source with some 100 other office workers was sent for ten days to help harvest at the Bolshaya Taskina Sovkhoz south of Lake Kaldy, about 50 kilometers north of Chelyabinsk. At the Nadyrov Bridge which crossed the Techa River, he saw a few posters with the inscription: "Drinking strictly prohibited, water polluted" (Pit strogo vospreshchayetsya, voda zagryaznena). While working at the sovkhoz, he did not approach the Techa, because the river bank was a prohibited area. Some distance north of the river there was a continuous ditch about one meter deep and one meter wide, with posters: "No passage, polluted zone" (Prokhod vospreshchayetsya, zagryaznennaya zona). Source did not discuss pollution of the Techa with persons on the sovkhoz. In Chelyabinsk he mentioned it to a friend, and was told that according to father who lived on the Techa somewhere in the Tyumen Oblast, the river was polluted on its lower course also.

6. Source vaguely remembered having heard that the Kyshtym area nuclear installation was known as the Post Box 40 installation. He knew that in 1960 the plant was managed by (fnu) Sorokin, whose daughter, Lyudmilla, born in 1930, source had met at a 1960 New Year's party in Chelyabinsk. She was a graduate of an institute in Sverdlovsk and was working at the Chelyabinsk town planning (gorproyekt) institute.

Exhibit 8. Excerpt from CIA report on nuclear installations in the USSR (dated February, 1961, and released January, 1977).

Exhibit 8 continued

7. In March 1958, an explosion wrecked part of the nuclear plant at Kyshtym. Whether the explosion was nuclear or chemical, source could not tell, nor did he have information on casualties. The matter was openly discussed among employees of the Urals Branch of the Academy of Construction and Architecture.

8. Source knew of one case in which work at the Kyshtym plant allegedly resulted in the sexual impotence of an engineer (name unknown) and subsequent divorce. The divorcee was Alina Loy (maiden name), an engineer with the trust Metallurgstroy at Chelyabinsk, who left her husband in 1956 or 1957 after a few months of married life. In summer 1960, she married (fnu) Chulkov, an officer with the combat engineers, who was transferred to Novaya Zemlya in August 1960.

9. While working at the Urals Branch of the Academy of Construction and Architecture, source heard that in 1957 its laboratory of reinforced-concrete construction (chief, /fnu/ Bershtein) had investigated an accident, fall of a smokestack from a huge plant which was being built by the MVD Glavpromstroy or Ministry of Medium Machine Building in the Argayash area.

10. Source was not certain but thought that a second plant might also have been built in the Argayash area by the MVD Glavpromstroy or Ministry of Medium Machine Building.

1. In late 1958 or early 1959, SOURCE heard of a restricted area which allegedly contained an "atomic plant". His vague description of one of the two possible locations placed it in immediate vicinity of the KYSHTYM.

Information based mainly on hearsay and rumor.

2. The only detail SOURCE could learn of the plant itself was that a "tube" protruded from the surface of a nearby lake. Area was tightly guarded by military personnel. Plant employees lived in a settlement within the restricted area, which they were not allowed to leave and which no unauthorized person could enter. In 1959, a female worker who died was buried within the

area; relatives were denied entry. Employees could send and receive mail. Address was CHELYABINSK Province, Post Office Box #40, plus street and name of addressee.

3. SOURCE heard that in spring 1959 a large, accidental explosion occurred. Many were killed, some of whom drowned following a landslide and resulting flood. Many others received overdoses of radiation and were evacuated to various towns in CHELYABINSK Province. Victims were subjected to regularly repeated medical examinations. Some time after the accident, SOURCE met a woman who had been affected by the radiation and had red, eczema-like markings on her face. As far as SOURCE knew, work was not discontinued as a result of the explosion. No further details.

EDITOR'S COMMENT: Although SOURCE never actually saw the restricted area, he was familiar with the surroundings and his information was gained from residents of the KYSHTYM area.

Exhibit 9. Declassified excerpts from CIA report on atomic accident in Kyshtym, 1958–1959.

CENTRAL INTELLIGENCE AGENCY
WASHINGTON, D.C. 20505

14 OCT 1977

Dr. Zhores A. Medvedev
National Institute for Medical Research
The Ridgeway, Mill Hill
London, England NW7 1AA

Dear Dr. Medvedev:

This will acknowledge your letter of 22 September requesting copies of CIA documents pertaining to the nuclear disaster in the Urals which you mentioned in your 4 November 1976 New Scientist article.

The Central Intelligence Agency is presently reviewing a number of documents relating to this event with a view to releasing as much information to the general public as possible. When this declassification review has been completed, we will be happy to furnish you with copies. Since some of these documents refer to your initial report, we will be happy to provide this service without cost.

Sincerely,

Gene F. Wilson
Information and Privacy Coordinator

Exhibit 10. CIA acknowledgment of author's letter of request for information.

THE NEW YORK TIMES, SATURDAY, NOVEMBER 26, 1977

C.I.A. Papers, Released to Nader, Tell of 2 Soviet Nuclear Accidents

By DAVID BURNHAM

Special to The New York Times

WASHINGTON, Nov. 25—The Central Intelligence Agency has made public 14 documents that describe two apparently separate nuclear accidents in the Soviet Union, one of which reportedly took the lives of hundreds of people.

The documents, made public in response to a Freedom of Information Act request by an antinuclear group established by Ralph Nader, appear to confirm a report of two nuclear accidents in the Soviet Union made public a year ago by Dr. Zhores A. Medvedev, an exiled Soviet scientist.

One of the C.I.A. documents, however, said it was possible that reports and memos of the nuclear accidents may have been prompted by a top-secret test in which the Soviet Union allegedly exploded a 20-megaton device in the air over a mock village populated with goats and sheep, to test the hazards of such an explosion.

Though most of the documents were anecdotal in form and considerable information had been deleted from them, it appeared that the two accidents occurred at a vast nuclear facility near the city of Kyshtym on the eastern slope of the Ural Mountains between 1958 and 1961. One of the reports, dated March 25, 1977, quoted an unnamed source as telling the C.I.A. that he had been told "hundreds of people perished and the area became and will remain radioactive for many years."

Affected Region Is Described

The source said that in 1961 he had visited the "strange, ininhabited and unfarmed area" where the accident reportedly had occurred. He described the region this way: "Highway signs along the way warned drivers not to stop for the next 20 to 30 kilometers because of radiation. The land was empty, there were no villages, no towns, no people, no cultivated land, only the chimneys of destroyed houses remained." Thirty kilometers would equal about 20 miles.

A former Soviet physicist, Leo Tumerman, who emigrated to Israel in 1972, described seeing virtually the same scene of desolation on an auto trip that he took through the Kyshtym area, in an account that appeared in the Dec. 9, 1976, issue of The New York Times. In it, Dr. Tumerman said he had been informed that he had passed through the site of the "Kyshtym catastrophe," named for a town in the vicinity, and that a nuclear disaster a few years earlier had killed and injured many hundreds of people. He said he thought the year of the explosion was in the late 1950's.

A second report, dated May 23, 1958, painted a portrait of a less serious nuclear accident: "Various Soviet employes and visitors to the Brussels fair have stated independently but consistently that the occurrence of an accidental atomic explosion during the spring of 1958 was widely known throughout the U.S.S.R." The Brussels World's Fair took place in 1958. The 1958 report added: "Rumors are common that many people were killed. However, the general accepted version is that only several score died."

Other reports described "a terrible explosion" that appeared to have occurred in either 1960 or 1961. The explosion was so great, the report said, that it made the ground and buildings shake. A short time after this explosion, it said, all the leaves on the trees in and around the blast area "were completely covered with a heavy layer of red dust."

Hospital Filled With Victims

This report said that a woman had been in a hospital "at the time of the explosion," and she said that after the blast occurred she saw many people, brought to this hospital for medical attention. The hospital was eventually filled with victims of the explosion.

Mr. Nader, in an interview, questioned the agency's motives in not making the documents public at an earlier date. "Absent any other reason for withholding information from the public," he said, "one possible motivation could have been the reluctance of the C.I.A. to highlight a nuclear accident in the U.S.S.R. that could cause concern among people living near nuclear facilities in the United States."

In November 1976 Dr. Medvedev, a dissident biochemist, writing in the British weekly New Scientist, charged that hundreds of people had been killed and thousands suffered radiation sickness in 1958 when atomic wastes buried in the Ural Mountains exploded.

Exhibit 11. *New York Times* on the Urals disaster.

The Washington Post

© 1977 Washington Post Co. SATURDAY, NOVEMBER 26, 1977 Phone (202) 223-6000 Classified 223-6200 Circulation 223-6100

CIA Data Confirm 2 Blasts at Soviet Atomic Site in '50s

By Bill Richards

Newly released U.S. intelligence documents show that two major explosions at a top-secret Soviet nuclear facility in the southern Ural Mountains killed or burned hundreds of persons in the late 1950s and left a large tract of land lifeless and contaminated with nuclear fallout.

It is not clear from the heavily censored Central Intelligence Agency files whether a 1958 blast at the Russian nuclear installation at Kyshtym was an atomic explosion. However, Soviet scientists apparently did set off a nuclear blast at the site during the following year which caused a wide fallout in the area.

A CIA informant traveling in the Kyshtym region in 1961 described the scene of the two blasts in these stark terms, according to the documents.

"We crossed a strange uninhabited and unfarmed area. Highway signs along the way warned drivers not to stop for the next 20 to 30 kilometers because of radiation. The land was empty. There were no villages, no towns, no people, no cultivated land, only the chimneys of destroyed houses remained.

"I asked the driver to stop because I wanted to drink water. The driver refused. 'One doesn't stop here. You drive quickly and cross the area without any stops,' he said."

The documents were obtained this week under a Freedom of Information

Exhibit 12. *Washington Post* on the Urals disaster.

Exhibit 12 continued

'50s Blasts at Soviet Atom Site Confirmed

BLAST, From A1

Act request by Ralph Nader's Critical Mass Energy Project and by The Washington Post. The CIA released 14 of the 29 documents it had on the incident and said some of the withheld reports were too sensitive to be released even with deletions.

The documents made available by the CIA confirm a report last year by exiled Soviet scientist Zhores Medvedev that thousand of persons were killed or suffered radiation sickness when buried nuclear waste at a site in the Urals overheated and exploded in 1958. Medvedev said the explosion released a radioactive cloud covering hundreds of miles in the area.

Other U.S. sources speculated last year that the explosion may have occurred at a site west of the Urals where they said it was believed plutonium-contaminated waste was stored in vats above ground. The sources said the vats may have been toppled by an earthquake.

The latest CIA documents make no mention of an earthquake. They note that the Kyahtym facility was closely guarded and closed to all outsiders except for a select group of Communist Party members and their dependents who were brought to the site from across the Soviet Union. The plant apparently manufactured components for nuclear weapons.

According to a CIA summary of information about the plant, hundreds of persons were exposed to radiation and injured in the 1958 explosion. In October of 1959 Soviet scientists apparently conducted an atomic test over the plant site contaminating food for at least 100 miles.

One CIA informant reported that after the 1958 blast, food purchased in Chelyabinsk, nearly 100 miles away from the plant site, was being carefully checked by authorities and in some cases destroyed because it was radioactive. The informant said some villages near Kyshtym were burned to the ground by authorities and the inhabitants evacuated with only the clothes on their backs

The CIA records note that the raw data supplied to the agency by its informants had not been evaluated for accuracy. Several informants' reports, for example, give different dates for the Kyahtym explosions than those contained in the CIA's own information summary.

The overall picture of the area following the blast is described by various informants, however, as one of devastation.

One report dated 1957-58 said all food stores in the region were shut and food was brought in two days later by truck and train. "The resulting queues were reminiscent of those during the worst shortages during World War II," the reports says. Some residents, it says, became hysterical with fear after the appearance of "mysterious" diseases while others walked around wearing radiation counters.

Another informant told the CIA in 1960 that after what was apparently the second blast in the area a fine red dust filtered down on the region around Chelyabinsk. "Very quickly," the informant said, "all the leaves curled up and fell off the trees."

The blast—which one CIA document indicated may have been a 20-megaton bomb deliberately set off over the site—was marked by a bright flash and a shaking of the ground in the area.

Victims of the nuclear explosion were treated in a local hospital where a wing was sealed off for them and all outsiders kept away. One informant at the hospital during that time reported "Some of them were bandaged and some were not. We could see the skin on their face, hands and other exposed parts of their body to be sloughing off . . . It was a horrible sight."

In an interview yesterday Nader said the blast was apparently the first serious nuclear accident involving massive casualties. Nader accused the CIA of deliberately holding up release of the documents about the blast for 20 years in order not to frighten persons in this country who might be concerned about nuclear development here.

"This information would have made people in the U.S. very concerned over nuclear installations, whether civilian or military," Nader said. He said also that he would seek to obtain at least some of the material withheld by the CIA on security grounds.

A CIA spokesman had no immediate comment yesterday on Nader's charges.

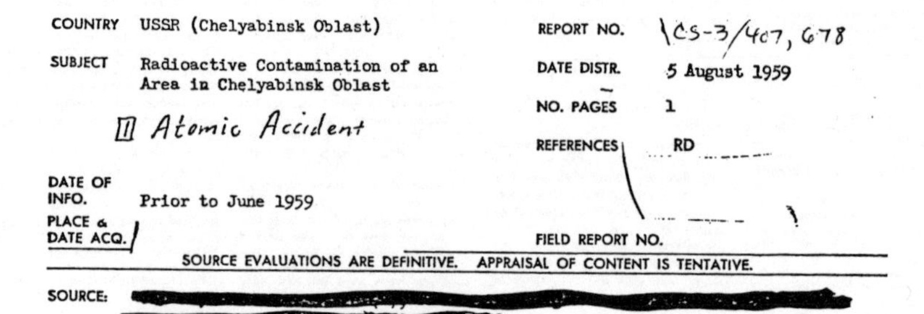

COUNTRY	USSR (Chelyabinsk Oblast)	REPORT NO.	CS-3/407,678
SUBJECT	Radioactive Contamination of an Area in Chelyabinsk Oblast	DATE DISTR.	5 August 1959
	II Atomic Accident	NO. PAGES	1
		REFERENCES	RD
DATE OF INFO.	Prior to June 1959		
PLACE & DATE ACQ.		FIELD REPORT NO.	

SOURCE EVALUATIONS ARE DEFINITIVE. APPRAISAL OF CONTENT IS TENTATIVE.

SOURCE:

On an unknown date prior to June 1959 an accident occurred in Chelyabinsk Oblast which caused radioactive contamination of the soil. As a result, the authorities were forced to take measures to resettle the inhabitants of several populated places in other areas.

Exhibit 13. "Sanitized" description of radioactive area. ("Document 8" of the set of 14 documents sent to author in response to request for information. Originally a full-page document, it had been almost completely "sanitized.")

DIRECTORATE OF OPERATIONS

Foreign Intelligence Information Report

COUNTRY	USSR	DCD REPORT NO.	OOK-323/20537-76
SUBJECT	Nuclear Explosion at Chelyabinsk-40/	DATE DISTR.	20 September 1976
		NO. PAGES	2
		REFERENCES	

3. According to the prevailing opinion in Chelyabinsk, Chelyabinsk-40 was a production site for nuclear devices. Chelyabinsk-40 is actually located in Kyshtym which is some one hundred kilometers northwest of Chelyabinsk. In about 1956 there was an explosion at Chelyabinsk-40; the explosion lighted up the sky for a great distance and the newspapers in Chelyabinsk made a flimsy attempt to proclaim the event an unusual occurrence of the northern lights. The chief evidence of the explosion was the tremendous number of casualties in the hospitals of Chelyabinsk. Many of the casualties were suffering from the effects of radiation. Shortly after the explosion a scientific research institute to study effects of radiation was established in Chelyabinsk, presumably as a result of the accident at Chelyabinsk-40. [Collector's comment: Source did not actually witness the explosion and could provide no details on its cause.]

APPROVED FOR RELEASE
Date 24 SEP 1977

Exhibit 14. "Sanitized" description of nuclear explosion. ("Document 11" of the set of 14 sent to author in response to request for information.)

DIRECTORATE OF OPERATIONS

Foreign Intelligence Information Report

COUNTRY	USSR	REPORT NO.	OOE-324/01015-77
SUBJECT	Soviet Detonation of 20 Megaton Device in 1950's in Above-Ground Test/ Possible Explanation for Recent News Reports on Nuclear Accident and "Vast Nothing" Area in Ural Mountains	DATE DISTR.	24 January 1977
		NO. PAGES	1
		REFERENCES	

DATE OF INFO. 1959 – 1960

THIS IS UNEVALUATED INFORMATION

SOURCE

1. Recently there have been accounts in US newspapers concerning comments made by two former citizens of the USSR on a "vast nothing", an area within the USSR where it is speculated a nuclear accident occurred in the late 1950's. There was a top secret Soviet film/ which showed a nuclear test that had occurred in an unspecified region of the Ural Mountains. It is likely, although not certain, that the test occurred in the 1957-58 period, and this may account for the "vast nothing" mentioned in the news accounts.

2. According to the film, the USSR constructed a completely new city in a valley in the Ural Mountains region for the test. A subway was constructed under the village, and one of the major purposes of the test was to see if the subway could withstand a nuclear attack. The inhabitants of the village were goats and sheep, and the post-explosion photography showed the effects of a nuclear blast upon animal life as well as building materials. Military equipment was placed around the village, and the effects of the explosion upon armaments of war also were depicted in the film.

3. The bomb itself was described as a 20 megaton device which was dropped from an airplane. The flash of the explosion illuminated the mountains which surrounded the village. The city virtually was eliminated, but the subway survived the explosion. Because of the film's classification, those who saw it were instructed to treat the whole matter as highly classified.

4. Recent newspaper accounts quote two Soviet emigres, one in London and the other in Israel, who knew something about the "vast nothing". One of the emigres said a 60 square mile area in the Ural Mountains was desolate and still heavily radioactive in 1961. It is possible the "nuclear accident" of which the emigres spoke is the event recorded by a Soviet camera crew and shown as a top secret defense film.

– end –

5 4 3 2 1

Caption at right-hand page

491;3

PLANT SUMMARY

1. Name

   Atomic Energy Installation

   [redacted]

2. Location

   KYSHTYM 55 44 N 60 35 E

   a. Address
      N/A

   b. Pinpoint
      The restricted area of Kyshtym is approximately 60 km N/S and 45 km E/W. The railroad from Kasli to Karabash runs diagonally from the NE corner to the SW corner with Kyshtym in the center. It includes the installations at Techa (reactor) and Sungul (radiological institute).

3. History

   A large atomic plant and a workers' settlement were established about 15 km NE of Kyshtym, probably at Techa on Osero Irtyash, during the period 1945 to 1948. Approximately 70,000 inmates of 12 labor camps, participated in the construction. In the spring of 1948, the entire population, including all PWs and forced laborers had to evacuate the Kyshtym restricted area. The population was replaced by Communists and their dependents who came to Kyshtym from all over the USSR. They were reportedly never to leave the area again.

4. Physical Plant and Equipment

   The restricted area covers 2700 sq. km containing eight small lakes with interconnecting watercources. The atomic plant (reactor) is situated in a tunnel which extends beneath a river, with only a smoke stack visible above ground. One of the lakes was drained and a building of undetermined size was built on its bed with cement, rubber, and lead. Then the lake was refilled with water. A double tracked RR line was built to the area. The underground factory was 30 to 40 meters below the surface and were as follows:
   8 small shops all the same size (approx. 50 by 25/28m). They had been blasted out from the slate rock. The vertical walls were coated with reinforced concrete up to a height of approx. 3 - 4 m. They supported a reinforced concrete three center arch roof 6 - 7 m high in the middle of the shop.
   The ceiling was more strongly armoured than the walls by the addition of cross bond iron bars.
   A large shop approx. 100 by 40 m was built in the same way as the smaller shops. The ceiling was supported in the middle of the room by 4 concrete pillars of 1,20 by 1,20 m cross section.

Exhibit 16. Description of an "ordinary" nuclear weapons test—in contrast to the Urals explosion (continued on p. 198).

Exhibit 15. Description of alleged 20-Megaton atomic explosion test. ("Document 13" of the set of 14 documents sent to author in response to request for information.)

Exhibit 16 continued

PLANT SUMMARY (Continued) IR Firm No. 8014401

4. Physical Plant and Equipment (Continued)

Construction of the shops was finished and the building of machine foundations started a few days before informant left the camp; these foundations were 1 by 2 m or 0,80 by 1,50 m.
All underground rooms were electrically lighted.
Steel brackets were cast in the walls of the large shop, which should possibly support the rails of cranes.
The thickness of the walls (consisting of the rock and concrete coating) between the shops was approx. 5 m.

5. Production

It is reported this plant contains atomic piles and supplies Sungul. Radiological Institute with radio-active materials. This plant has been reported to be manufacturing components for atomic weapons
In the spring of 1958 hundreds of persons were exposed to radiation and injured as a result of an explosion at the Kyshtym plant.
In early October 1959, an atomic test reportedly took place in Kyshtym. After the test, such foodstuffs as meat, fish and milk were removed from the retail stores in Sverdlovsk and Chelyabinsk and destroyed. Residents were ordered to turn in food stocks in their houses. Residents were warned against buying agricultural products from farmers.

6. Labor

In this area in 1956 there were military personnel from various army units and arms. With them 16 labor battalions of about 1,000 men each were activated. There were also 25,000 Soviet soldiers of General Vlasov, who had collaborated with the Germans. These men were actually considered as prisoners and were likewise organized into labor battalions. In addition, about 60,000 Soviet convicts of both sexes were employed in the project.

7. Key Personalities and Organization

N/A

8. Security

Strict security observed. Movement was restricted in the vicinity of the plant. The surrounding fences were considerably removed from the enterprise itself, but the entrances were under permanent military guards. Special passes required.

9. Visits by American and/or Western Observers

N/A

10. Photos Available

N/A

BIBLIOGRAPHY AND NOTES

Bibliography and Notes

(1) Medvedev, Zhores A. "Two Decades of Dissidence," *New Scientist*, vol. 72, no. 1025 (1976), 264–67.

(2) Medvedev, Zhores A. "Facts Behind the Soviet Nuclear Disaster," *New Scientist*, vol. 74, no. 1058 (1977), 761–64.

(3) Khrushchev, Nikita S. *Khrushchev Remembers: The Last Testament*. Boston-Toronto: Little, Brown & Co., 1974. Chap. 18.

(4, 5, 6) *Sbornik rabot Laboratorii biofiziki Ural'skogo filiala Akademii Nauk SSSR* (Collected Papers of the Biophysics Laboratory of the Urals Branch of the USSR Academy of Sciences). Vol. 1, Sverdlovsk, 1957. Vol. 2, Moscow, 1960. Vol. 3, Sverdlovsk, 1962.

(7) Timofeeva-Resovskaya, E. A. *Raspredelenie radioizotopov po osnovnym komponentam presnovodnykh vodoemov* (The Distribution of Radioisotopes Among the Basic Components of Fresh-Water Systems). Sverdlovsk: The Press of the Urals Branch of the USSR Academy of Sciences, 1962.

(8) Timofeev-Resovsky, N. V. *Nekotorye problemy radiatsionnoi biogeotsenologii. Doklad po opublikovannym rabotam, predstavlennym dlia zashchity uchenoi stepeni doktora biologicheskikh nauk*

(Some Problems of Radiational Biogeocenology: A Report on [His] Published Works Submitted for the Degree of Doctor of Biological Sciences). Sverdlovsk: The Press of the Urals Branch of the USSR Academy of Sciences, 1962.

(9) Rovinsky, F. Ya "Sposob rascheta kontsentratsii radioaktivnoi primesi v vode i donnom sloe neprotochnykh vodoemov" (Calculation Method for the Distribution of Radioactive Contamination in the Water and Bottom Deposits of Non-Running-Water Lakes), *Atomnaia energiia* (Atomic Energy), vol. 18, no. 4 (1965) 379–83.

(10) Ilenko, A. I. "Nakoplenie strontsiia-90 i tseziia-137 presnovodnymi rybami" (Accumulation of Strontium-90 and Cesium-137 by Freshwater Fish), *Voprosy ikhtiologii* (Problems of Ichthyology), vol. 10, no. 6 (1970), 1127–28.

(11) Ilenko, A. I. "Nekotorye osobennosti nakopleniia tseziia-137 v populiatsiiakh ryb presnovodnogo vodoema" (Some Peculiarities of Cesium-137 Accumulation in the Fish Populations of a Freshwater Lake), *Voprosy ikhtiologii,* vol. 12, no. 1 (1972), 174–78.

(12) Ilenko, A. I. *Kontsentrirovanie zhivotnymi radioizotopov i ikh vliianie na populiatsiiu* (The Concentration of Radioisotopes in Animals and Their Effect on the Animal Population). Moscow: "Nauka" (Science) Press, 1974.

(13) Ilenko, A. I. "Radioekologiia presnovodnykh ryb" (Radioecology of Freshwater Fish), *Voprosy ikhtiologii,* vol. 9, no. 2 (1969), 324–37.

(14) Parkhomenko, G. M. "Predel'nye dopustimye dozy oblucheniia i kontsentratsiia radioaktivnykh veshchestv v vozdukhe i vode" (Maximum Permissible Radiation Doses and Concentrations of Radioactive Substances in Air and Water), in *Radiatsionnaia gigiena* (Radiation Hygiene), edited by M. S. Gorodinsky. Moscow: Meditsina Press, 1962. P. 31.

(15) Koz'min, A. "O rybokhoziaistvennom ispol'zovanii ural'skikh vodokhranilishch" (The Use of Urals Lakes for Commercial Fishing), in *Biologicheskaia produktivnost' vodoemov Sibiri* (Biological Productivity of Siberian Lakes), edited by M. Iu. Bekman. Moscow: Nauka Press, 1969. Pp. 203–8.

(16) Lindel, B., and R. L. Dobson. *Ioniziruiushchaia radiatsiia i zdorov'e* (Ionizing Radiation and Health). A publication of the World Health Organization, Geneva, 1961. (The World Health Organization also publishes this material in English.)

(17) Turner, F. B., B. Kowalewsky, R. H. Rowland, and K. H. Larson. "Uptake of Radioactive Materials from a Nuclear Reactor by Small

Mammals at the Nevada Test Site," *Health Physics*, vol. 10 (1964) 65–68.

(18) Martin, W. E., and F. B. Turner. "Transfer of Sr-90 from Plants to Rabbits in a Fallout Field," *Health Physics*, vol. 12 (1966), 621–31.

(19) Svensson, G. K., and K. Liden. "The Transport of Cs-137 from Lichen to Animal and Man," *Health Physics*, vol. 11 (1965), 1393–1400.

(20) Ilenko, A. I., and G. N. Romanov. "Sezonnye i vozrastnye izmeneniia moshchnostei doz v skelete temnykh polevok ot inkorporirovannogo strontsiia-90 v estestvennykh usloviiakh" (Seasonal and Age-related Changes in the Dosage Strength of Strontium-90 Incorporated under Natural Conditions in the Skeletons of Dark Field Mice), *Radiobiologiia* (Radiobiology), vol. 7, no. 1 (1967), 76–78.

(21) Ilenko, A. I. "Faktory, opredeliaiushchie uroven' nakopleniia radioaktivnogo strontsiia-90 v populiatsii temnykh polevok, obitaiushchikh na zagriaznennoi territorii" (Factors Determining the Level of Accumulation of Radioactive Strontium-90 in a Population of Dark Field Mice Inhabiting a Contaminated Territory), in the collection *Ekologiia ptits i mlekopitaiushchikh* (The Ecology of Birds and Mammals). Moscow: Nauka Press, 1967. Pp. 126–32.

(22) Ilenko, A. I. "Materialy po vidovym razlichiiam v nakoplenii strontsiia-90 i izmenchivosti melkikh mlekopitaiushchikh, otlovlennykh na uchastke iskusstvenno zagriaznennom etim radionuklidom" (Materials on Species Differences in Sr^{90} Accumulation and Variability in Small Mammals Taken in a Field Artificially Contaminated by This Radionuclide), *Zoologicheskii zhurnal* (Journal of Zoology), vol. 47, no. 11 (1968), 1695–1700.

(23) Sokolov, V. E., and A. I. Ilenko. "Radioekologiia nazemnykh pozvonochnykh zhivotnykh" (Radioecology of Terrestrial Vertebrates), *Uspekhi sovremennoi biologii* (Advances in Contemporary Biology), vol. 67, no. 2 (1969), 235–55.

(24) Nikitina, N. A. "Itogi izucheniia peremeshchenii gryzunov fauny SSSR" (Results of a Study of Migration in Rodent Species of the USSR), *Zoologicheskii zhurnal*, vol. 50, no. 3 (1971), 408–26¡

(25) Lishin, O. V. "Materialy po dinamike chisla polevok" (Material on Numerical Changes in Field Mice), in *Trudy Komissii po okhrane prirody* (Proceedings of the Commission on the Conservation of Nature), issue 1, Urals Branch of the USSR Academy of Sciences, Sverdlovsky, 1964. Pp. 183–86.

(26) Ilenko, A. I., and E. A. Fedorov. "Nakoplenie radioaktivnogo tseziia v populiatsiiakh nazemnykh pozvonochnykh" (Accumulation of Radioactive Cesium in Terrestrial Vertebrate Populations), *Zoologicheskii zhurnal,* vol. 49, no. 9 (1970), 1370–76.
(27) Ilenko, A. I. "Zakonomernosti migratsii strontsiia-90 i tseziia-137 v raznykh zven'iakh pishchevykh tsepei v zootsenoze" (Migration Patterns of Sr^{90} and Cs^{137} at Various Food-Chain Levels in a Zoocenosis), *Zhurnal obshchei biologii* (Journal of General Biology), vol. 31, no. 6 (1970), 698–708.
(28) Korsakov, Iu. D., I. Ia. Popylko, and I. A. Ternovsky. "Organizatsiia dozimetricheskogo kontrolia vneshnei sredy pri radiatsionnykh avariiakh" (The Organization of Dosimetric Monitoring of the External Environment in Radiation Accidents), in the book *Handling of Radiation Accidents:* Proceedings of the Symposium on the Handling of Radiation Accidents Sponsored by the International Atomic Energy Agency (IAEA), Vienna, 19–23 May, 1969. Vienna: IAEA, 1969. Pp. 281–85.
(29) Ilenko, A. I., and A. D. Pokarzhevsky. "Vliianie biotsenoticheskikh razlichii na kontsentrirovanie strontsiia-90 melkimi mlekopitaiushchimi" (The Effect of Biocenotic Differences on Sr^{90} Concentration in Small Mammals), *Zoologicheskii zhurnal,* vol. 51, no. 8 (1972), 1219–24.
(30) Naumov, N. P. *Ekologiia zhivotnykh* (Animal Ecology). Moscow: Vysshaia Shkola (Higher School) Press, 1963.
(31) Shvarts, S. S. *Evoliutsionnaia ekologiia zhivotnykh. Trudy instituta ekologii rastenii i zhivotnykh Ural'skogo filiala Akademii Nauk SSSR* (Evolutionary Ecology of Animals: Proceedings of the Institute of Plant and Animal Ecology of the Urals Branch of the USSR Academy of Sciences), issue 65. Sverdlovsk, 1968.
(32) *Lesa Urala i zemledelie* (The Forests of the Urals and Agriculture). Sverdlovsk, 1968.
(33) Ilenko, A. I., S. I. Isaev, and I. A. Riabtsev. "Radiochuvstvitel'nost' nekotorykh melkikh mlekopitaiushchikh i vozmozhnost' adaptatsii populiatsii gryzunov k iskusstvennomu zagriazneniiu biogeotsenoza strontsiem-90" (Radiosensitivity in Several Small Mammals and the Possibility of Adaptation in Rodent Populations to Artificial Contamination of a Biogeocenosis by Sr^{90}), *Radiobiologiia,* vol. 14, no. 4 (1974), 572–75.
(34) Ilenko, A. I. "Nekotorye zakonomernosti kontsentratsii iskusstvennykh radioaktivnykh izotopov ptitsami lesnogo biotsenoza" (Some Regularities in the Concentration of Artificially Introduced Radio-

isotopes in the Birds of a Forest Biocenosis), *Zoologicheskii zhurnal,* vol. 49, no. 12 (1970), 1884–86.

(35) Voous, Karel Hendrick. *Atlas of European Birds.* N.p., 1960.

(36) Dement'ev, G. P., and N. A. Gladkov. *Ptitsy Sovetskogo Soiuza* (Birds of the Soviet Union), vols. 1–6. Moscow, 1951–1954.

(37) Ilenko, A. I., and I. A. Riabtsev. "O gnezdovom konservatizme nekotorykh vodoplavaiushchikh ptits" (On Nesting Conservatism in Several Species of Water Birds), *Zoologicheskii zhurnal,* vol. 53, no. 2 (1974), 308–10.

(38) Ilenko, A. I., I. A. Riabtsev, and D. E. Fedorov. "Izuchenie territorial'nogo konservatizma otkrytognezdishchikhsia vorob'inykh ptits metodom radioaktivnogo mecheniia populiatskii" (A Study of Territorial Conservatism in Open-Nesting Passerines by the Radioactive Tagging Method), *Zoologicheskii zhurnal,* vol. 54, no. 11 (1975), 1678–86.

(39) Willard, W. K. "Avian Uptake of Fission Products from an Area Contaminated by Low-level Atomic Waste," Science, vol. 132 (1960), 148–50.

(40) Aleksakhin, R. M., and F. I. Pavlotskaia. "Migratsiia radionuklidov v pochvakh i rasteniiakh. Simpozium v Tbilisi" (The Migration of Radionuclides in Soil and Plants: A Symposium in Tbilisi), *Vestnik AN SSSR* (Bulletin of the USSR Academy of Sciences), 1971, no. 7, 123–26.

(41) Giliarov, M. S., and D. A. Krivolutsky. "Radioekologicheskie issledovaniia v pochvennoi zoologii" (Radioecological Research in Soil Zoology), *Zoologicheskii zhurnal,* vol. 50, no. 3 (1971), 329–42.

(42) Krivolutsky, D. A., A. L. Tikhomirova, and V. A. Turkhaninova. "Structuranderungen des Tierbesatzens (Land- und Bodenwirbellose) unter dem Einfluss der Kontamination des Bodens mit Sr^{90}" (Structural changes in Animal Populations [Surface and Soil Invertebrates] Affected by Contamination of the Soil with Sr^{90}), *Pedobiologia,* vol. 12, no. 8 (1972), 374–80.

(43) Ilenko, A. I. "Zarazhennost' melkikh mlekopitaiushchikh gazovymi kleshchami na uchastkakh, zagriaznennykh strontsiem-90" (Infestation of Small Mammals with Gamasid Mites in Areas Contaminated by Sr^{90}), *Zoologicheskii zhurnal,* vol. 50, no. 2 (1971), 234–46.

(44) Krivolutsky, D. A., and A. D. Pokarzhevsky. "Rol' pochvennykh zhivotnykh v biogennoi migratsii kal'tsiia i strontsiia-90" (The Role of Soil Animals in the Biogenic Migration of Calcium and Sr^{90}), *Zhurnal obshchei biologii,* vol. 35, no. 2 (1974), 263–69.

(45) Krivolutsky, D. A., and S. A. Shilova. *Vestnik Moskovskogo Gosudarstvennogo Universiteta* (Bulletin of Moscow State University), series 5, Geography, no. 3 (1965), 72–75.

(46) Krivolutsky, D. A., and A. F. Baranov. "Vliianie radioaktivnogo zagriazneniia pochvy na naselenie murav'ev" (Effect of Radioactive Soil Contamination on an Ant Population), *Zoologicheskii zhurnal,* vol. 51, no. 8 (1972), 1248–51.

(47) Kornberg, E. I., M. I. Dziuba, and V. I. Zhukov. "Areal kleshcha *Ixodus ricinus* v SSSR" (Habitat of the Mite *Ixodus ricinus* in the USSR), *Zoologicheskii zhurnal,* vol. 50, no. 1 (1971), 41–48.

(48) Sparrow, A. N., and G. M. Woodwell. "Prediction of the sensitivity of plants to chronic gamma irradiation," *Radiation Botany,* vol. 2 (1962), 9–26.

(49) Dubinin, N. P., V. A. Shevchenko, A. Ia. Alekseenok, L. V. Cherezhanova, and E. M. Tishchenko. "O geneticheskikh protsessakh v populiatsiiakh podvergaiushchikhsia khronicheskomu vozdeistviiu ioniziruiushchei radiatsii" (Genetic Processes in Populations Subjected to the Chronic Effects of Ionizing Radiation), *Uspekhi sovremennoi genetiki* (Advances in Contemporary Genetics), an annual review edited by Dubinin, vol. 4. Moscow: Nauka Press, 1972. Pp. 170–205.

(50) Tikhomirov, F. A. *Deistvie ioniziruiushchikh izluchenii na ekologicheskie sistemy* (The Effect of Ionizing Radiation on Ecological Systems). Moscow: Atomizdat (Atom Press), 1972.

(51) Dubinin, N. P. *Vechnoe dvizhenie* (Eternal Motion). Moscow: Gospolitizdat (State Publishers for Political Literature), 1973.

(52) Aleksakhin, R. M., M. A. Naryshkin, and M. A. Bocharova. "K voprosu ob osobennostiakh i kolichestvennom prognozirovanii kumuliativnogo nakopleniia strontsiia-90 v drevesnykh nasazhdeniiakh," *Doklady Akademii Nauk SSSR* (Reports of the USSR Academy of Sciences), vol. 193, no. 5 (1970), 1192–94.

(53) Aleksakhin, R. M., and M. A. Naryshkin. *Migratsiia radionuklidov v lesnykh biogeotsenozakh* (The Migration of Radionuclides in Forest Biogeocenoses). Moscow: Nauka Press, 1977.

(54) Makheev, A. K., and S. A. Mamaev. *Ekologiia* (Ecology), 1972, no. 1, 24–36.

(55) Aleksakhin, R. M., F. A. Tikhomirov, and N. A. Kulikov. "Sostoianie i problemy lesnoi radioekologii" (The Status of Forest Radioecology and Its Problems), *Ekologiia,* 1970, no. 1, 19–26.

(56) Tikhomirov, F. A., R. A. Aleksakhim, and E. A. Fedorov. "Migratsiia radionuklidov v lesakh i deistvie ioniziruiushchikh izluchenii na

lesnye nasazhdeniia" (The Migration of Radionuclides in Forests and the Effect of Ionizing Radiation on Forest Stands), in *Peaceful Uses of Atomic Energy: Proceedings of the 4th International Conference on the Peaceful Uses of Atomic Energy, Sept. 6–16, 1971.* Vol. 11. Vienna: IAEA, 1972. Pp. 675–88.

(57) Klechkovsky, V. M., and E. A. Fedorov. [Thesis of a Report], in the book *Migratsiia radioaktivnykh elementov v nazemnykh biotsenozakh* (The Migration of Radioactive Elements in Terrestrial Biocenoses). Moscow: Nauka Press, 1968. P. 25.

(58) Cherezhanova, L. V., R. M. Aleksakhin, and E. G. Smirnov. "O tsitogeneticheskoi adaptatsii rastenii pri khronicheskom vozdeistvii ioniziruiushchei radiatsii" (Cytogenetic Adaptation of Plants Affected by Chronic Ionizing Radiation), *Genetika* (Genetics), vol. 7, no. 4 (1971), 30–37.

(59) Cherezhanova, L. V., and R. M. Aleksakhin. "K voprosu o tsitogeneticheskom vliianii mnogoletnego vozdeistviia povyshennogo fona radiatsii na populiastsii rastenii v prirodnykh usloviiakh" (The Cytogentic Effect of Many Years of Heightened Radiation Background on Plant Populations in Natural Conditions), *Zhurnal obshchei biologii,* vol. 32, no. 4 (1971), 494–500.

(60) Cherezhanova, L. V., and R. M. Aleksakhin. "O biologicheskom deistvii povyshennogo fona ioniziruiushchikh izluchenii v protsessakh radioadaptatsii v populiatsiiakh travianistykh rastenii" (Biological Effect of Heightened Ionizing-Radiation Background on Radioadaptation Processes among Herbaceous Plant Populations), *Zhurnal obshchei biologii*, vol. 36, no. 2 (1975), 303–11.

(61) Shevchenko, V. A., L. V. Cherezhanova, and A. Ia. Alekseenok. "Uvelichenie radiorezistentnosti v prirodnykh populiatsiiakh nizshikh i vysshikh rastenii pri dlitel'nom vozdeistvii beta-izlucheniia $Sr^{90} + Y^{90}$" (Increased Radioresistance in Natural Populations of Lower and Higher Plants under the Prolonged Effect of Beta Radiation from Sr^{90} with Y^{90}), in *Materialy I Vsesoiuznogo simpoziuma po radiobiologii rastitel'nogo organizma* (Materials of the 1st All-Union Symposium on the Radiobiology of the Plant Organism). Kiev: Naukova Dumka (Scientific Thought) Press, 1970. P139.

(62) Shevchenko, V. A. "O geneticheskoi adaptatsii populiatsii khlorealy k khronicheskomu vozdeistviiu ioniziruiushchei radiatsii" (On the Genetic Adaptation of Chlorella Populations to the Chronic Effect of Ionizing Radiation), *Genetika,* vol. 6, no. 8 (1970) 64–73.

(63) Fuller, John G. *We Almost Lost Detroit.* New York: Ballantine, 1976.

(64) Patterson, Walter C. *Nuclear Power*. London: Penguin, 1976.
(65) Ryle, Martin. "Nuclear Energy: The Serious Doubts That Put Our Future at Risk," *The Times* (London), December 14, 1976.
(66) See: U.S. AEC, PNE-201F (1962), and U.S. AEC, PNE-217F (1963).
(67) Golovanov, I. N. *I. V. Kurchatov*. Moscow: Atomizdat, 1967.
(68) Astashenkov, P. *Kurchatov*. Moscow: Molodaia Gvardia (Young Guard) Publishers, 1967.
(69) Gillette, R. "Radiation Spill at Hanford: The Anatomy of an Accident," *Science*, vol. 193 (1973), 728–30.
(70) Nuclear Reactor Safety: Hearing before the Joint Commission on Atomic Energy of the Congress of the United States. 93d Congress, Part I. Washington, D.C.: U.S. Government Printing Office, 1974.
(71) Report on the Investigation of the 106 T Tank Leak at the Hanford Reservation, Richland, Washington. U.S. Atomic Energy Commission, 1973.
(72) WASH-1520. Environmental Statement: Contaminated Soil Removal Facility, Richland, Washington, April 1972. U.S. Atomic Energy Commission.
(73) *Los Angeles Times Magazine*, June 18, 1972, pp. 5–7.
(74) Hill, Sir. John. "Letter to the Editor," *The Times* (London), February 8, 1977.
(75) Accident at Windscale No. 1 Pile on 10th October, 1957. Report Presented to Parliament by the Prime Minister by Command of Her Majesty, November 1957. London: Her Majesty's Stationery Office, 1957.
(76) Belitzky, Boris. "Removing Radioactive Rubbish in the USSR," *New Scientist*, vol. 69 (1976), No. 989, 436–37.
(77) Belitzky, Boris. "The Soviet Answer to Nuclear Waste," *New Scientist*, vol. 74 (1977), No. 1048, 128–29.
(78) Marsily, G. de, E. Ledoux, A. Barbreau, and J. Margat. "Nuclear Waste Disposal: Can the Geologist Guarantee Isolation?" *Science*, vol. 197, no. 4303 (1977), 519–27.
(79) Krugmann, H., and F. von Hippel. "Radioactive Wastes: A Comparison of U.S. Military and Civilian Inventories," *Science*, vol. 197 (1977), 883–85.
(80) Parker, H. M. "Radioactive Waste Management in Selected Foreign Countries," *Nuclear Technology*, vol. 24 (1974), 307.
(81) Angino, E. E. "High-level and Long-lived Radioactive Waste Disposal," *Science*, vol. 198, no. 4320 (1977), 885–88.

Index